The Climatic Threat

John Gribbin was born in 1946. He took a BSc in Physics and an MSc in Astronomy at the University of Sussex, followed by a PhD in Astrophysics at the University of Cambridge. For five years he worked on the staff of *Nature*, and for most of that time was in charge of a daily science report for *The Times*. Since 1975 he has been a freelance writer/broadcaster, and has been engaged in research at the University of Sussex into the problems of climatic change. Among his other books are: *Our Changing Climate*, *Galaxy Formation*, *Forecasts, Famines and Freezes*, *Climatic Change*, *White Holes*, *Our Changing Universe* and, with Stephen Plagemann, *The Jupiter Effect*.

John Gribbin

The Climatic Threat

What's wrong with our weather?

Fontana/Collins

First published in Fontana 1978

Made and printed in Great Britain by
William Collins Sons and Co Ltd, Glasgow

To all my friends at SPRU, with thanks

Contents

Shall we feed those who walk to our tune, those who are friendly? Shall we feed only the nations that control their population growth? Shall we feed only the nations that have something of value to trade to us, or only the nations that guarantee freedoms we hold dear? Or shall we simply feed the nations that are hungry?

WALTER ORR ROBERTS 1976

Preface

The problems of climatic change have been brought home to everyone so clearly by the events of recent years that there is surely no need for me to explain here my chief motive in writing this book – in order to provide information about how and why the climate is changing for as wide an audience as possible, and to indicate the best available forecasts of the changes in climate we can expect to experience in the next few years and decades. But there is a secondary reason why I should have written this particular book at this particular time.

As an astronomer who came to study climatic change and its impact on human society, I was initially wary of the various 'astronomical' theories of climatic change, and in particular of the ideas of a link between solar (sunspot) activity and the weather on Earth. There are very many ideas about why the climate changes, and in seeking to obtain a balanced view of the whole picture I deliberately suppressed my natural instinct to favour astronomical ideas. One result of this balanced approach was my previous book on climate, *Forecasts, Famines and Freezes*, in which I attempted to sum up the present understanding of climate and climatic change for other non-climatologists who were becoming interested in these problems.

But, having tried to balance the pros and cons of different theories, and having suppressed any instinctive feelings that the astronomical ideas might be better than some of the other theories, I found a curious change in my attitude developing. The more I looked at the other theories, the more I became convinced that the links between the Earth and Sun really could provide the best explanation of much of the decade to decade, and century to century, variation in climate that is of key importance to humankind. When climatologists at the US National Center for Atmospheric Research – scientists with no background in astronomy – began to point out how well the 'sunspot' theory could explain these climatic changes, and took astronomers to task for being unable to answer the questions about the Sun raised by studies of the Earth's changing climate, I decided the time had come to throw caution to the winds, return to my astronomical roots, and put together a more subjective story concentrating on the links between Sun and Earth and their impact on climatic change and human society. This book tells that story.

For anyone seeking a cool appraisal of climatic change, a balanced view with all theories spoken for, put this book down and read *Forecasts, Famines and Freezes* (or, at a more technical level, the volume *Climatic Change* which I edited for the Cambridge University Press). But if you are intrigued by the idea of changes in the Sun affecting our weather and climate, and through them our daily lives, read on. What you will find is a subjective interpretation of the evidence, made by someone who has come to believe that this is the single best explanation of what has gone 'wrong' with the weather lately. You may find that you disagree with my interpretation, and want to know more about the theories which are given less attention here – that's OK, you can always read my other books afterwards! Whatever theory of climatic change you subscribe to, however, one thing remains clear. The Earth's climate is indeed changing, and is increasing the strain imposed on agriculture by the growing needs of a growing global population of hungry people. We must learn to understand and live with the threat of climatic change, and I hope that this book will contribute to that understanding.

JOHN GRIBBIN, May 1977

Chapter One

What's wrong with the weather – and where is it going?

There is now little doubt that the world's climate is changing, and has changed, away from the benign conditions of the middle part of this century. There is even less doubt that the world food supply and distribution systems are in a crisis state, with no more than one month's supply in reserve, insufficient to compensate for even one bad harvest that might result from the natural fluctuations of weather and climate. This situation makes an understanding of the changing climate, and a reliable prediction of its future pattern, more urgent than ever before – and this is the view not just of a few climatologists seeking support for their own work, but of respected government organizations and agencies.

A US National Research Council report published in 1976 summed up the situation in the following terms:

> An understanding of climatic change, both natural and man-made, is one of the central issues society must address . . . In 1976, food for 4 billion inhabitants is required; in the year 2000, food for at least 6 billion will be needed.

Also in 1976, a CIA working paper on 'A Study of Climatological Research as it Pertains to Intelligence Problems', written three years earlier still, was declassified and made publicly available. Warnings such as the comment that 'With global climatic-induced agricultural failures in the early 1970s, the stability of many governments has been seriously threatened' received little attention then. But when the US was hit simultaneously by severe drought in the west and blizzards in the east a few months later, in the winter of 1976–7, suddenly the CIA-sponsored work became a focus of attention.

In Europe too, where a few voices had been warning of the climatic threat for years, the topic became respectable when record-breaking droughts in some regions were followed by record-breaking floods. In January 1976, news agencies carried reports of a prediction from West German meteorologists that Europe's north-west coast could be facing a new 'storm age', with a greatly increased threat of flooding and prospects of delays and difficulties with production in the newly developed North Sea oilfields. All the evidence points to a new climatic era of extreme variations of climate, from place to place and from season to season, covering literally the entire globe.

At the South Pole, 1976 brought the lowest temperatures ever recorded since the South Pole Base was established in 1957 – an average daily temperature of −50·0°C, with the coldest day being 8 August, reaching a bracing −76·0°C. For two seasons in succession, scientists at the British research base at Halley Bay in Antarctica could not be properly resupplied because of the refusal of the pack ice nearby to break up sufficiently for the supply vessel to reach the base, and for months there was a real prospect that the base would have to be abandoned. Some supplies did get through early in 1977, but the deteriorating climate continues to leave a big question mark over the future of the base. In the North, an item in *Soviet News* reported a study of climatic change in the Arctic, made by Soviet scientists who believe that 'the Arctic is slowly and steadily becoming colder and the North Pole area is reverting to the harsh conditions which prevailed during the last century . . . it is anticipated that the Arctic will continue to grow colder at least until the year 2000'.

In between those geographical extremes, the climate showed the same pattern of extreme conditions almost everywhere. Table 1.1 shows a country-by-country listing of exceptional climatic events in 1976, taken from the review of the year's climate published in the *Climate Monitor* of the Climatic Research Unit at Britain's University of East Anglia. The same publication draws attention to the probable cause of these extremes, referring to the work of the Director of that Unit:

Professor Hubert Lamb has in recent years frequently drawn

attention to the increasing frequency of occurrence of stationary (or 'blocking') situations in the circumpolar air current of the northern hemisphere. The persistence of such blocks produces in middle latitudes long spells of persistent weather which modify the condition of the Earth's surface – parching or soaking the ground over wide regions, warming or cooling the seas – and so, encourage the development of extremes of temperature and of the moisture regime.

Table 1.1 Country by country listing of exceptional climatic events in 1976

Australia: Drought in Victoria and parts of southern Australia in July after floods in February.

Austria: Severe gales associated with N. European depression on 3 February. Excessive rains in late summer and early snow in September.

Azores Islands: Hurricanes 'Emmy' and 'Frances' strike in early August.

Bangladesh: Drought from January to May followed by unusually heavy 'burst' of monsoon (Sylhet: 18 inches in 24 hours on 9 June) causing floods and landslides. Monsoons were preceded in April and May by tornadoes with wind speeds estimated at 150 mph.

Belgium: Drought from January to August. Heatwave in late June.

Benin, People's Republic of: Exceptional long-lasting rainfall (October–November) in northern regions, shortfall of second rainy season (October–December) precipitation in equatorial regions.

Brazil: Worst drought for ten years in central Brazil ended in early February but continued until September in north-eastern Brazil.

Burma: Bay of Bengal storm causing unseasonal rainfall at the end of December in Lower Burma.

Canada: Wettest summer on record in British Columbia.

Chile: Severe drought during July and August preceded by extreme winds (140 km/h) in June and followed by excessive rains in October.

Costa Rica: Severe drought in Quanacaste Province with estimated $13 million loss in basic grains production.

Finland: Temperatures over Lapland fell to $-40°C$ during the second half of December.

Germany, Federal Republic of: Drought from the beginning of the year, most severe in northern areas. Lowest rainfall and highest number of successive days with extreme maximum temperatures since records have been kept.

Hong Kong: Typhoon 'Ellen' caused flooding in August.

India: Temporary interruption and early end of monsoon rains affected food production adversely. Flooding in Madras, Tamil Nadu and Andhra Pradesh in late November. Cyclone hit Gujarat state in early June; 40,000 homeless. Serious flooding in northern states and in Assam during July and August.

Indonesia: January–April: floods and destructive winds. May: destructive winds. June–September: drought and destructive winds. October: destructive winds. November–December: floods and destructive winds.

Iran: Floods in July.

Israel: Exceptionally high incidence of frost resulting in damage to many crops. Shortage of winter rains in southern Israel caused moisture stress on crops.

Italy: Torrential rains in February, July, August, October and November.

Japan: Typhoons 'Billy' (August) and 'Fran' (September) caused extensive damage, especially in Okinawa.

Kenya: Drought in southern region (Ukambani). Flash floods causing train disaster in November.

Korea, Republic of: Severe shortfall of rain during winter 1975/76. Serious flooding during August and September.

Liberia: Exceptionally heavy rainfall on 13 September which paralysed Monrovia and caused much damage.

Madagascar: Three tropical cyclones of exceptional intensity.

CLOTILDE (10–14 January) north-western coast, wind speeds in excess of 200 km/h;

DANAE (21 January) Majunga to north-eastern coast;

GLADYS (March) north-east (Prefecture d'Antalalia).

Total damage: 16 deaths, 26 injured, 8850 homeless, $40,000 material.

Malawi: Excessive rainfall (maximum in December) caused unprecedented high levels of Lake Malawi.

Malaysia: Severe drought and bush fires during first half of the year.

Mali: Unusually high rainfall in late October.

Mexico: Hurricane 'Liza' struck Baja California on 1 October with unusually high intensity causing the death of 1000 inhabitants.

Nepal: Interruption of the early monsoon caused 15–20% reduction in rice production. Record low temperatures in December.

Netherlands: Drought from the beginning of the year until end of August. Heatwaves in May, June and July.

New Zealand: First snowfall since 1904 in northern areas.

Niger: Unprecedented high rainfall in October.

Norway: Heaviest snowfall on record in January. All glaciers in West and North Norway increasing.

Oman: Abnormally high rainfall during period January–April.

Pakistan: Excessively heavy rainfall in July, August and September caused widespread devastations due to flooding in the Indus catchment, overflowing and eventual collapse of the Bolan dam.

Paraguay: Drought during second half of the year.

Sahel: Severe rainfall deficiencies during June–August in certain areas followed by unseasonal rainfall in September–October.

Sierra Leone: Drought until September followed by unusually heavy rainfall in October.

South Africa: Excessive rainfall in March: Durban flooded.

Sri Lanka: Severe drought from the beginning of the year, which reached maximum intensity from May to September, reduced tea, coconut and rice production and caused acute water shortage in Colombo.

Sudan: Failure of rains in August followed by excessive rain in November (usually dry).

Swaziland: Two tornadoes (12 and 18 November) causing at least ten deaths and considerable material damage.

Thailand: Excessive monsoon rainfall and flash floods during 20–30 November and on 12 September.

Turkey: Severe winter with snowstorms and avalanches causing many casualties in January and February.

United Kingdom: Severe gales associated with exceptionally deep depression on 2/3 January. Drought from the beginning of the year to end of August. No rain at all fell during April in Callington, Cornwall; 38 successive dry days were recorded at Oxford. Heat-waves in early May, June and July. Mean temperatures between mid-June and 8 July probably higher than in any period of the same length since 1719. Heavy rains during September and October provided the wettest autumn in the 250-year rainfall record. 11 September was wettest day recorded since records began in 1852 at Durham (87·8 mm), and since 1868 at Glasgow (87·3 mm). September rainfall total for Durham was 381·8% of long-period average. December brought very cold conditions with extensive snowfall over Scotland and parts of England.

Upper Volta: Failure of rains in south-eastern region during July and August. Exceptionally heavy rains in October and early November.

USA: Extensive and prolonged drought since 1975 in Pacific seaboard areas, notably California and some mid-western states. Flash floods in Colorado during August. Hurricane 'Belle' caused extensive damage on eastern seaboard in August.

USSR: Apart from local exceptions, generally favourable crop weather producing a near-record grain harvest. Severe winter with excessive snow in southern regions (Republic of Georgia) in January. Excessive rainfall during mid-summer in Central European Russia caused severe flooding of Moscow in August. Severe cold spell in October and November, especially in Kazakhstan.

Venezuela: Floods in mid-July.

Yugoslavia: Heavy rains from 6–8 June caused flooding of 25,000 hectares of land in southern Siberia.

From *Climate Monitor* review of 1976 (edited by P. P. Jones).

The European pattern

'Extremes of temperature and of the moisture regime' – a perfect description of what hit Europe and America in 1976–7, but perhaps not very descriptive in everyday human terms. Before we have a closer look at the causes and consequences of this 'blocking' weather situation, it seems a good idea to look in a little more detail at the nature of the climatic extremes we have been living through. Each case, taken alone, might not be regarded as exceptional – one swallow doesn't make a summer. But put the whole together and the picture is grim indeed.

First, the new storm age. Even before the rigours of 1976–7, storms in the North Sea had caused a setback in North Sea oil development which dealt a seven-figure blow to the UK balance of payments, holding up production from British Petroleum's Forties Field for two months. This was only a minor hitch – but the incredible cost to the economy of the country of such a small hitch in production from one field for such a short time indicates just how sensitive is our dependence on oil today. In the spring of 1977, a blow-out in a well in the Norwegian sector of the North Sea occurred, causing extensive pollution as well as loss of production, and here again news reports told of the difficulties caused to workers tackling the problem by 'unseasonable' storms.

In the winter of 1975–6, the Scottish tourist industry was hit when snow failed to arrive at ski resorts in the Cairngorms; just a year later there was plenty of snow, but still problems for the tourists who found not just flights but road and rail transport north severely hampered, with Scotland actually completely cut off from England (except by sea) on one memorable day.

But to European citizens it was the summer of '76 that really brought home the implications of the climatic threat. The effects of the drought wrecked the farm budget of the EEC, with no better example than that of the continuing problem of Europe's over-production of dairy products. It might seem that a dry summer would help Europe's administrators in their attempts to get rid of the vast butter and dairy 'mountain' that has resulted from deficiencies in past policies, since there would be less production. In fact, however, the drought coincided with attempts to provide a long-term solution to the dairy problem by changing subsidies and imposing a 'milk tax'. Farmers complained that such action would be the last straw in a desperate year, resulting in bankruptcy and no milk production at all. The result? No firm, permanent action to control the dairy mountain, and stop-gap solutions boosting the European Commission's supplementary budget way beyond £500 million.

The European harvest was hit by the drought to the tune of about 11 million tonnes of cereal crops with the 1976 yield of 91·4 million tonnes comparing almost disastrously with the 1971–5 average of 103 million tonnes. In the UK alone, drought losses were estimated by the Deputy Director-General of the National Farmers' Union as cutting farm incomes by up to £400 million, a reduction of 40 per cent across the board. And all this at a time when just a little way to the east, in global terms, unprecedented heavy rains in 1976 were producing record-breaking crops in parts of the USSR, although handicapping the farmers in their harvesting endeavours.

Of course, the great drought brought its lighter moments, not least being a sudden demand for package holidays to Siberia, booked by sun-tanned Britons desperate to get away from it all. Holidaymakers who stayed at home had their problems in spite of the blue skies and blazing sun which should have made holiday-making ideal. Hundreds of miles of Britain's canal system, a popular tourist attraction, were closed for the duration of the drought, and one news story warned drivers heading for the West Country of Devon and Cornwall that they would have to take water with them – not for themselves, but for the radiators of their thirsty vehicles, which could no longer be topped up at water-rationed garages in the holiday centres.

One of the most curious stories of the effect of a global climatic change on every aspect of our daily lives appeared in the pages of the rock weekly *New Musical Express*, concerned because the world is running out of blue denim suitable for making jeans. Droughts hitting cotton harvests caused the shortage of denim, and this coincided with a shortage of blue dye, with the giant Imperial Chemical Industries group announcing that they could not justify the investment needed to increase production because of the uncertain nature of the fashion business. Of course, the shortage of cloth won't matter too much if we have more hot summers – we can all switch to shorts instead!

But Britain simply is not equipped to cope with extremes of

weather. And much more sober statistics also show the problems of continued extremes of any kind. In Greater London, the death rate rose by more than 10% during the heatwave, with the number dying in the week ending 2 July, for example, reaching 11,536 compared with a usual weekly average of about 10,000. Declining levels in reservoirs across England and Wales and the slow flow of drying-up rivers increased the concentration of nitrates in drinking water above the accepted safety levels in some regions, where mothers of babies under six months old were advised to mix their infants' feeds using bottled water supplied by local medical authorities. And back in London again a problem familiar to inhabitants of Los Angeles or Tokyo, but not in benign old England, reared its head – photochemical smog produced by the action of all that sunlight on exhaust material from motor vehicles. The action taken by some car workers in striking because temperatures in their workshop reached an intolerable 100°F did not last long enough to affect that problem, but highlighted the fact, not noticed in the years we have come to think of as 'normal', that in the UK although the law sets down a required minimum temperature for working conditions in offices, shops and factories there is no set upper limit defined as tolerable for working in.

North American contrasts

All this happened in the summer of '76, and the struggle of Europeans, and the English in particular, to cope with such conditions was received with some amusement in the US. A few months later, however, the smile was quickly wiped away when North America suffered almost exactly the same persistent climatic pattern – drought in the west, heavy precipitation in the east – with the difference that with the pattern becoming set in mid-winter it was the precipitation, falling as snow and with record low temperatures, that made the biggest items of headline news. In fact, in the longer term view the western drought is likely to have even more impact on the economy of the US.

In much of the US this drought was but one more in a succession of dry years, with some farmers in Nebraska recalling the 'dust bowl' period of the 1930s as being *less* severe than the present drought. Governor J. Exxon of Nebraska was reported by newsman Ian Ball as estimating crop losses amounting to $400 million, with 30,000 families affected by the drought. The arrival of winter did nothing to help. Farmers not just in Nebraska but across Kansas, North and South Dakota, Oklahoma, California, Texas, Wyoming and Colorado were all, as of spring 1977, expecting a severe impact on crops in the current year, with the lack of snow in the past winter exacerbating conditions that have been building up for two or three years.

The irony of this problem was that in the east, both in Canada and the US, they had far too much snow in the winter of 1976–7. Millions of workers were laid off; dozens of people died in the snow; and the economy received a hammer blow of a $6000 million dollar bill for the damage, with growth in Gross National Product (GNP) of the US reduced by one per cent to 4–4½% in the first quarter of 1977 and to 4·8% instead of 5·1% for the full year, according to US Chamber of Commerce chief economist Jack Carlson. The impetus given to new President Jimmy Carter in his efforts to cut back on the growth in consumption of energy by the US is now history, as are the stories of the winter that have already become part of legend and folklore . . .

Drought hit Colorado loaning snowploughs to snowbound Buffalo, New York . . .

Polar bears in Alaska refusing to hibernate as the same blocking pattern that brought Ice Age conditions to the eastern US brought them warm air from the Pacific . . .

Snow being loaded on to railroad freight cars in New York State, for transportation south where it could melt without bringing a flood problem . . .

The threat of an inter-state legal suit when Wayne Kidwell, chief legal officer of parched Idaho, complained that cloud-seeding operations in next door Washington State amounted to 'stealing' Idaho's rain . . .
and so on.

At the annual meeting of the American Association for the Advancement of Science, held in Denver, Colorado in February 1977, the topsy-turvy weather, and especially the drought, inevitably became the topics of the week, and the report of the meeting carried in *Science* described how a minor session scheduled for a small conference room to cover the topic 'American Droughts' suddenly became jam-packed with both scientists and journalists. Dr Stephen Schneider, Deputy Head of the climate project at the National Center for Atmospheric Research in nearby Boulder soon dispelled any lingering illusions held by those present that the bizarre weather conditions were unpredicted, pointing out that a long series of highly publicized scientific warnings foreshadowed the severe weather conditions, but that nothing had been done to prepare for such disasters. There was a failure to stockpile food or launch water conservation projects, no attempt to increase natural gas supplies, and no real effort, until the winter brought home the warnings, to produce an effective energy conservation programme.

The problem with the climatic predictions, as Schneider acknowledged, was that they could not be precise. The climatologists had warned of the increasing risk of extreme conditions – but they could not put their finger on the calendar and say 'February 1977 is going to bring record snow in the eastern US'. But this shouldn't stop preparation for increased risks – the actuarial approach of the insurance business, which can take account of the probability of death and disaster without knowing precisely when and where it will strike. The problem with our present society is that we lack insurance, in the form of food reserves or slack in the energy supply system, and elsewhere.

From one extreme to another

This unpredictability of weather in the short term itself becomes greater when the climate shifts into a pattern of extreme variations. At the end of August 1976, forecasters in England saw no end to the drought; days later the rains started and continued almost unabated for two months. The driest sixteen-month period in the records of England and Wales was followed by the second wettest September–October period since records began. The farmers, initially smiling through the rain while the met. men blushed, soon faced a new problem. What little in the way of sugar beet and potatoes that had survived the drought recovered well in the rain, but could only be harvested with extreme difficulty in flooded fields. Much of the crop rotted where it lay. Cattle which now had some grass to eat quickly found their fields disappearing below water; and, one of the greatest ironies, because of the delay needed to fill up reservoirs even with torrential rainfall, many towns and villages in the west of England had main water supplies cut off, with standpipes brought into use for restricted supplies on street corners, at a time when more than 70 mm of rain – two-thirds of the normal monthly total for the region in September – fell in the space of five days. Table 1.2 and Fig. 1.1 highlight this dramatic turnabout – but there was more to come to bemuse the battered inhabitants of the British Isles. After record drought, record flood and record heatwave, what else should they have expected but record cold and snow? And that is just what they got.

Of course, a hard winter by English standards isn't quite the same as a record-breaking snowfall in Siberia, or even upstate New York. But the arrival of the heaviest snows experienced in all but one of the previous thirty years did round out the astonishing saga of 1976–7 and leave everyone wondering what on Earth to expect next. It wasn't just the heaviest snow for years over most of the country, it was the *first* real snow for years in much of the south, where many young drivers experienced snowy roads for the first time, and wished they hadn't.

So both sides of the Atlantic were hit by extremes of weather – and both sides of the Pacific, too. In February 1977, Japan's northern regions suffered the heaviest snowfalls for fourteen years, and a month later news leaked out from China concerning severe water shortages affecting agricultural areas along the Huai and Yellow rivers. The climatic threat is very much something that is affecting the whole world – but to see what is happening in a historical context, and to obtain that forecast for the future that is so badly needed, we need to focus our attention

Table 1.2 Shown below are the monthly areal general rainfall totals for England and Wales covering the period January 1970 to the latest available record. Entries are in whole millimetres. The standard 35-year (1916–1950) monthly averages are also shown for comparison purposes.

Year	*Jan*	*Feb*	*Mar*	*Apr*	*May*	*Jun*	*Jul*	*Aug*	*Sep*	*Oct*	*Nov*	*Dec*	*Annual*
1970	106	82	64	86	26	46	71	79	65	56	175	56	912
1971	109	33	66	54	51	104	43	109	26	71	94	38	798
1972	103	73	79	67	73	76	59	38	45	32	99	104	848
1973	44	40	24	67	83	63	92	63	86	57	52	68	739
1974	117	98	47	14	40	66	77	95	144	99	125	72	994
1975	117	31	81	71	47	21	66	52	107	36	73	50	752
1976	60	40	43	21	64	17	32	32	160	153	83	94	799
(Average 1916–1950)	92	66	57	60	63	55	79	81	76	92	95	88	904

Source: UK Meteorological Office.

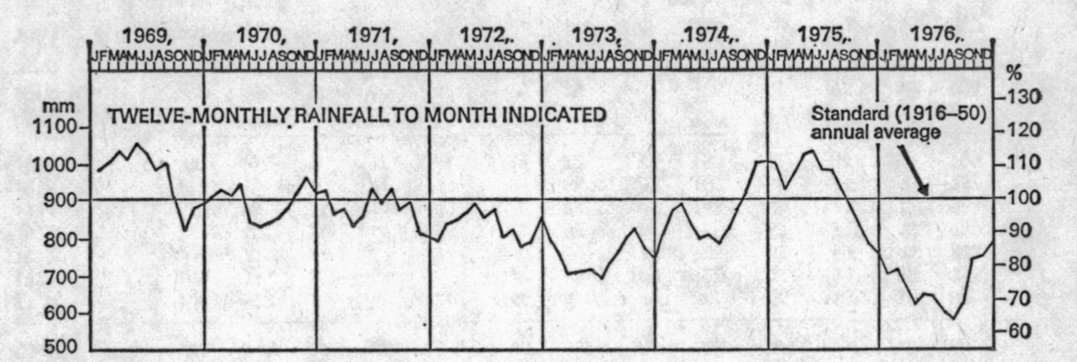

Figure 1.1 A longer perspective indicates that the drought in England and Wales in 1976 was no sudden freak, but simply the driest year yet in a rainfall decline which set in at the end of the 1960s.

From *Climate Monitor* review of 1976 (edited by P. D. Jones).

on a more limited geographical region.

Ideally, climatologists would like to settle on North America as the region to study, since after all this is the region where what little surplus food the world has is produced today. But records of climatic history from the region are hardly very helpful. The winter of 1976–7, for example, has now gone down as the most severe, with the coldest January, in the entire history of the United States. But that history only goes back for two centuries – and the pre-history of colonial days scarcely any further! In climatic terms, that is but the blink of an eye, and we need longer records from which to develop an understanding.

As we have seen, though, the climate of the whole globe is what matters, and it does form one pattern. It is no coincidence that the worst US winter of the past two centuries came so quickly on the heels of the most severe English drought of the past two centuries. So the way to develop an understanding of the way the climate is changing, and a prediction of what this will mean for all of us, is first to get hold of the broad picture of how changes in Europe relate to changes around the globe, then to use the available records from Europe, and especially England, to get a grasp of how climate has changed over the past centuries. In this way, we can fill in the future pattern world-wide – not in any sense of providing detailed predictions of what next summer or next winter will be like, but in the sense of indicating what chance there is of more droughts, or floods, or more heatwaves, or more severe winters. And the key to all this is that circulation pattern mentioned at the very beginning of this chapter – the 'blocking' circulation discussed by Professor Lamb.

Freak weather? Or back to 'normal'?

The underlying theme of many popular reports of the bizarre conditions of 1976–7 was that they were 'just one of those things' (or perhaps just several of those things?) that happen from time to time. Even the respectable authority of the UK Meteorological Office held the official view that since climate does vary, and weather varies every year, sometimes the variations will just be more extreme than others, and on this picture even, say, the English drought of 1976 must be something that will happen sooner or later. We were just unlucky to be the people who lived through it, so the official line runs. Plausible enough put like that, but not so plausible when you remember that in the same period North Americans were 'just unlucky' in their weather; the Chinese were 'just unlucky' with their droughts; the Japanese were 'just unlucky' with severe snow; and in the five years of the 1970s leading up to these events inhabitants of Africa, India, South America and Australia had also been 'just unlucky' with droughts, floods, frosts and other climatic events unparalleled in living memory.

Clearly, when put in the correct perspective these events tell us that the whole global pattern of climate has shifted. What we are used to thinking of as 'normal' – the climate of the thirty years up to 1970 – just is not 'normal' any more. And, an even more sobering discovery, the best evidence from studies of available historical records is that in the longer perspective it is those benign years in the middle part of the twentieth century that look unusual. In the longer term, it seems that the weather pattern that has hit us in the 1970s actually marks a return to normal in terms of world climate!

Dr R. A. S. Ratcliffe (who, as it happens, does work at the UK Meteorological Office) mentioned one feature of the drought/flood pattern of England's weather in 1976 that supports this view of a major climatic shift. From the late nineteenth century up to the 1970s, he reported in an article in *Weather*, fine summers have usually heralded fine Septembers. But a fine summer in 1975 was followed by a wet September, and the record-breaking drought of 1976 was followed by record-breaking rain in the September/October period. This is just the pattern we find when we look back in the records beyond 1727 – in the decades before 1727 we can find thirteen particularly fine, dry summers, none of which was followed by a dry September. And the same story of a return to conditions of two or three centuries ago emerges from more detailed study of the rainfall records.

In January 1977, the Royal Meteorological Society held a

meeting in London to discuss 'The Drought of 1975–6'. Among the various contributions was a report by J. M. Craddock, of the University of East Anglia, which put the whole business in its proper historical perspective. Craddock turned the usual comforting argument of the meteorological establishment on its head. Where others contended that such a drought was so unlike normal conditions that we could safely bet it wouldn't happen again in our lifetimes, he pointed out that when you have a theory of the way the weather works, but something happens that is far outside the limits of what the theory regards as 'reasonable', then the time has come to change your theory. The drought of 1975–6 may have been a real freak compared with, say, 1940–70; but coming after a run of dry years in the 1970s (see Fig. 1.1), and being so extreme, the drought is best interpreted as a harbinger of the new climatic pattern. And, again, it is in the records of the eighteenth century that a precedent for such events can be found. As Craddock puts it, referring to the data shown here in Fig. 1.2, 'the decline of the rainfall totals from 1965 onwards has an ominous resemblance to that which occurred between 1730 and 1740, which brought rainfall down to about the same level, and was followed, during the next 20 years, by two more years with rainfall below 16 inches'. This is no proof that two more severe European droughts are due in the next couple of decades, but at least hints that the wetter conditions of recent decades are not a reliable long-term guide to 'normal' weather conditions.

If only the Native Americans had been as interested in rainfall as the English, and had the written records in which to report weather changes and hand them on to succeeding generations, we would be in a much better position to see the global importance of this kind of variation of weather over the centuries. As things stand, though, we must rely on the English and other European records – the only comparable source of historical weather data is China, and that is not a very accessible source today.

So, to make the most of the data we do have on rainfall, take a look at Fig. 1.3, where Craddock has averaged out the rainfall

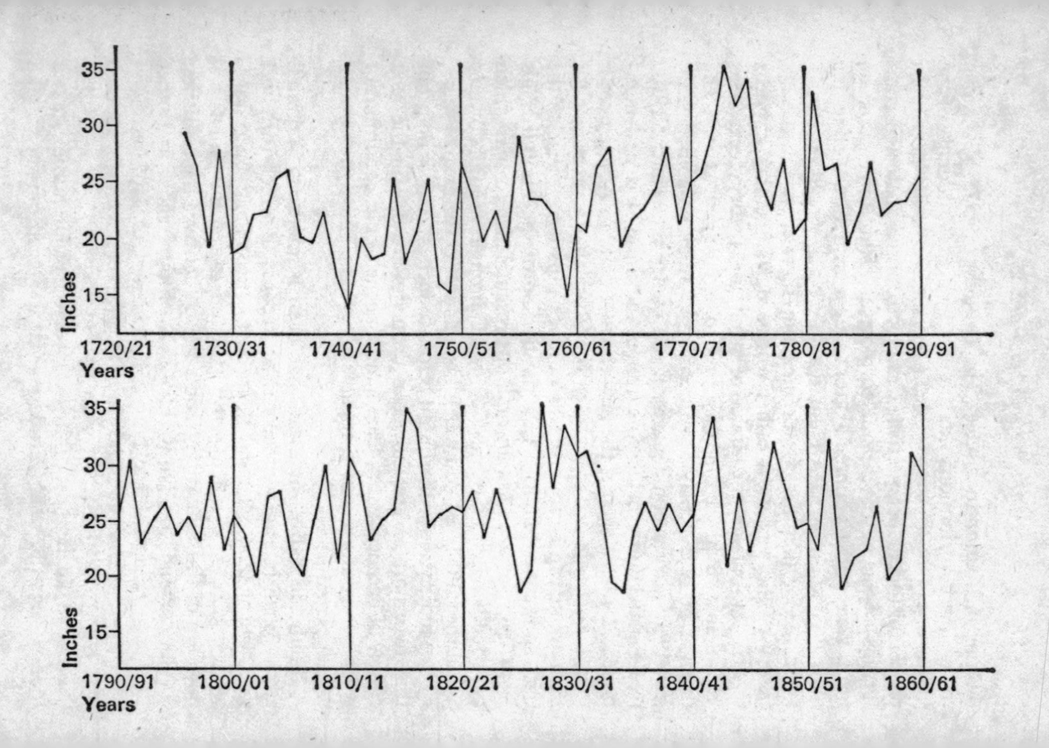
35
30
25
20
15
Inches
1720/21
1730/31
1740/41
1750/51
1760/61
1770/71
1780/81
1790/91
Years
35
30
25
20
15
Inches
1790/91
1800/01
1810/11
1820/21
1830/31
1840/41
1850/51
1860/61
Years

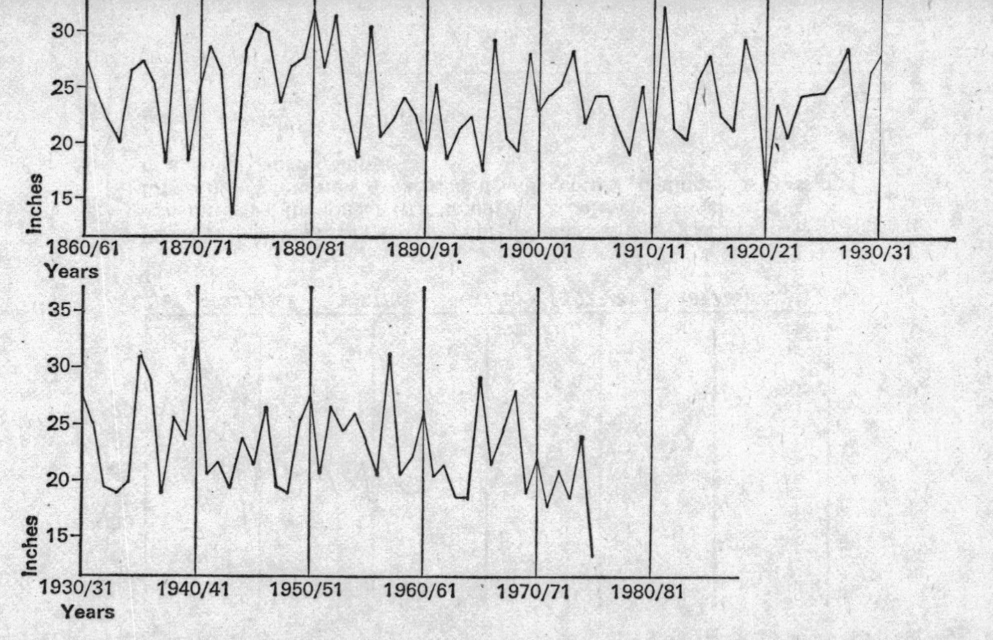

Figure 1.2 Annual rainfall totals (for twelve monthly periods ending in August of the indicated year) from 1726/27 to 1975/76, for the Pode Hole site in central England.

Figure supplied by J. M. Craddock.

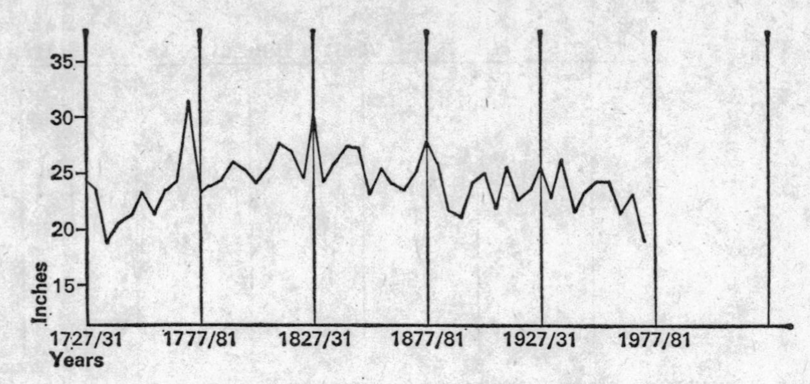

Figure 1.3 When the Pode Hall rainfall figures are averaged over five-year intervals, the longer term trends are seen even more clearly; is England now returning to early eighteenth-century conditions, as part of a global climatic shift?

Figure supplied by J. M. Craddock.

figures in successive five-year intervals. And look again at what he has to say about the implications:

> This shows a consistent, and perhaps a rather disturbing, picture of the rainfall variations in the Pode Hole area during the past 250 years. From a dry minimum in 1732/6, of below 20 inches, the rainfall rose to an average of around 26 inches in about 1827/31, and since then has fallen away, to perhaps 23 inches in about the 1960s, and to below 20 inches in 1972/6.

Within this longer-term pattern we have the even worse news for farmers that present conditions do not spread what little rain there is conveniently through the year, but bring us droughts followed by floods. What can any agricultural planner make of the detailed breakdown of month by month rainfall indicated by the figures of Fig. 1.1 and Table 1.2? How can you plan ahead when your dried-up crops are then flooded by conditions which make harvesting impossible? And all this is just taking the example of one small country. What we are in for really is a climate of extremes, and studies of the atmospheric circulation pattern can explain the how, although not yet the why, of such a dramatic change of conditions.

Weak circulation and the blocking pattern

The circulation changes, and their effects on climate in the middle latitudes of the Northern Hemisphere in particular, are most simply understood by looking at the flow pattern of the jet stream. This great river of air sweeps around the Earth high above the ground, about six miles above sea level, with windspeeds ranging from scores to hundreds of knots. For most of the twentieth century, these winds blew almost in a circle around the polar regions – the jet stream contained few 'wiggles', and the so-called 'strong' circulation pattern brought a consistent flow of weather systems at sea level from west to east underneath the jet stream. In England, this meant a succession of low pressure systems bringing rain from the Atlantic; since the oceans are cooler than the land in summer and warmer in winter, this westerly flow also brought even temperatures and few dramatic variations of any kind.

In North America, the weather was also relatively predictable, although the barrier of the Rockies could upset the even flow from time to time, contributing to such disasters as the great drought of the 1930s. Strong circulation, we know from historical climatology, goes hand in hand with even-tempered weather and a slight warming of the globe. But all that has now changed.

The globe has cooled down slightly in recent decades, and this has brought a shift into a so-called 'weak' circulation pattern. This means that the jet stream zig-zags much more widely as it snakes around the globe, like a wriggling snake chasing its own tail. Where the wind swings up towards the polar regions, blowing from south-west to north-east, it brings warm air northwards – as happened to Alaska in the winter of 1976–7. Then, where the stream sweeps around and back down to the south-east, it carries freezing air and plenty of snow down south – as the eastern seaboard of the US down to Florida experienced in that same winter (see Fig. 1.4). Even that wouldn't matter too much, unless the circulation system stayed stuck in the same groove for weeks at a time – but that is just what has been happening lately.

We can get a rough idea of why this should be so by looking down on the whole pattern from above the North Pole, as indicated by Fig. 1.5. If the jet stream 'snake' continues to wiggle, with the head failing to catch the tail, we will get rapid changes from one extreme to another, as weather patterns coming up from the south-west are replaced by a pattern bringing weather down from the north-west. But when the number of wiggles in the jet is just right, the head of the snake catches its tail and the pattern stabilizes into a standing wave – exactly equivalent to the vibrations of a plucked guitar string, or the sound waves vibrating in an organ pipe. Now, the snake ceases to thrash, the pattern stabilizes and the whole circulation gets locked in to a blocking state. The long wave pattern can exist with exactly or nearly 2, 3, 4, 5 or 6 wiggles in it, and although in principle it can get 'locked in' to any of these states, it is the higher numbers which bring more extreme conditions, with weather coming from either more

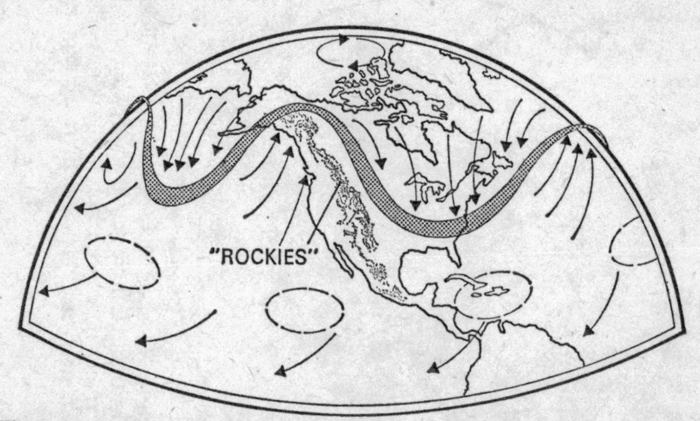

Figure 1.4 The wiggle in the jet stream that brought the Arctic US winter of 1977 is indicated schematically in this diagram. In the weak circulation state of the atmosphere, pronounced bends in the jet stream which guides weather systems at ground level can occur. In this case, the high altitude jet stream pushed way north into Alaska (bringing a warm winter) before diving southwards around the Rockies, carrying Arctic air into the eastern seaboard of the US. With weather systems carried northwards away from the west coast, this pattern also explains the drought experienced in many states.

northerly or more southerly directions, depending on just where in the world you are. The pattern that caused all the trouble in 1975–7 was wave number 5; and as the circulation remains weak there is every chance of more droughts (and floods) in Europe, and of more droughts in the western US and more cold winters in the eastern US.

In the case of the English drought we have followed in such detail, it is now easy to see what happened. The jet stream curling north around Britain carried sea-level rain systems with it along path C in Fig. 1.5, sweeping down into the Baltic region and dumping the 'English' rain on Scandinavia, Moscow and Leningrad. For some reason not yet explained, in the fall this pattern 'flipped' into the alternative state indicated by path E in Fig. 1.5, so that now weather systems from the North Atlantic

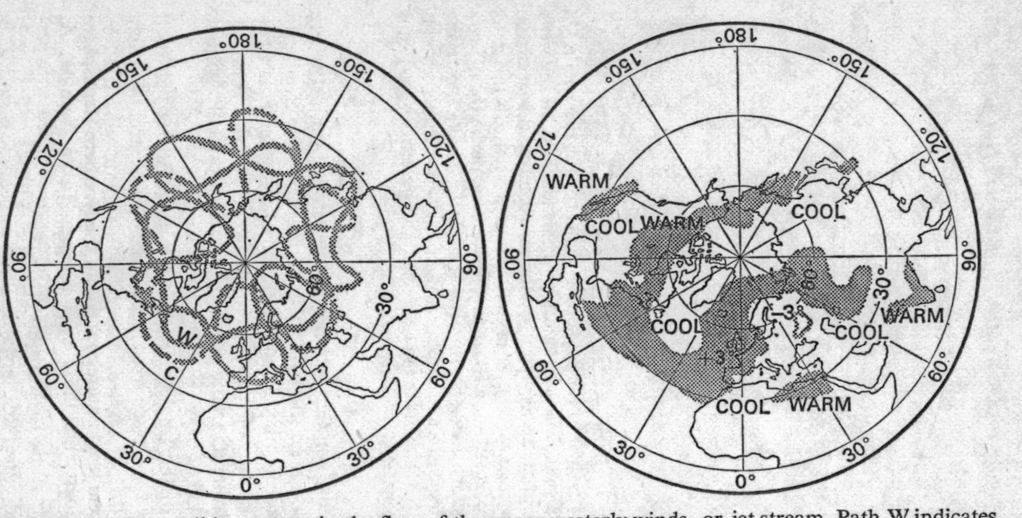

Figure 1.5 left, possible patterns in the flow of the upper westerly winds, or jet stream. Path W indicates a 'strong' circulation pattern with regular westerly winds; paths C and E indicate blocking patterns, with weak circulation, typical of recent years. Compare with Fig. 1.4. right, consequences of the blocking pattern are clearly seen in this reconstruction of temperature patterns in summer 1976. Deviations are expressed relative to the long term mean.

Figures supplied by H. H. Lamb.

swept into England and then became stuck, marking time until they could turn around to follow the lead of the jet stream again. Of course, this is a highly simplified picture of what went on. Not every low pressure system exactly follows the path marked out by the jet stream – all of the above description applies to some hypothetical 'average' state. What is surprising, in view of this great oversimplification, is just how good an insight this simple picture of the behaviour of strong and weak circulation provides into the mysteries of recent dramatic weather patterns.

What we have learned, even from the simplistic picture, is that a cooling of the globe brings 'weak' circulation, a predisposition towards blocking, and a climate of extremes. We have also seen that recent events are more like those of 200 or more years ago than like those of 20 years ago. To the obvious first question this raises – is the world in fact cooling down? – we need only the evidence of Fig. 1.6 and Table 2.1 (see p. 55) to assure us that the middle of the twentieth century was indeed unusually warm, as

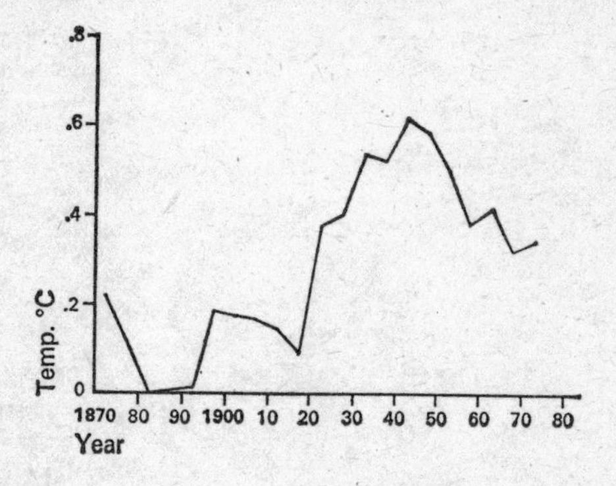

Figure 1.6 Recent temperature changes for the Northern Hemisphere showing deviations from the 1880–4 average. Taken from W. A. Brinkman's paper in *Quaternary Research*, volume 6, p 355, with permission of the author and the publisher. The outstanding feature is that the conditions we think of as 'normal' in fact marked an unusual temperature peak.

well as unusually even-tempered in climate, when viewed in the perspective of recent centuries. So, armed with this simple overview of just what has gone wrong with the weather we are ready to look in more detail at just what the true historical perspective on climate is. And then, perhaps, we might begin to tackle the jackpot questions – what is causing these changes, and what do they mean for our climatic future?

Chapter Two

The lessons of history

The realization that the climate of the world can and does change, rapidly enough and by a large enough amount to affect our present way of life, is the first and most important lesson of climatic history. The recent events outlined in Chapter One have brought this lesson home only too clearly yet until very recently climatic 'experts' could still be found supporting the view that climate is simply some kind of 'average weather', and that if you accumulate records of changes in the weather over a few decades you will have reliable averages indicating how climate should behave over the decades to come. Now, we can see that there is no reason to expect that the 'average weather' for the period from 1950 to 1999, say, will turn out to be the same as the average from 1900 to 1949. But just how different might the weather of the coming decades be from the averages over the earlier part of the twentieth century, which we regard as 'normal' since they describe the most recent patterns of the Earth's changing climate? If we understood exactly how the changes in climate are brought about, and if we had perfect information about the state of the atmosphere and oceans today, we could in principle predict what is going to happen in great detail. But this is far from being the case, and our present understanding of climatic processes, and the forces which drive them, is far from complete. We shall see in later chapters how one group of theories, in particular, provides a good explanation of recent climatic changes, and therefore offers a fair means of predicting future patterns of climate. But with the present-day imperfection of understanding it is vital to draw as much information as possible from the records of past climates, in the expectation that the extreme variations of the near future will probably be no more extreme than the most extreme weather patterns of recent history.

The questions are, firstly, how far ahead do we wish to 'predict' and, secondly, how far back need we go to get a reliable guide to possible extremes of climate? As far as predicting future patterns goes, there is probably little point in trying to get even a sketchy outline for more than a century ahead, both because our understanding of climate and ability at forecasting is likely to improve dramatically within a hundred years, and also because the increasing influence of man's activities seems likely to change the natural patterns of climatic change to a noticeable degree by the middle of the twenty-first century (see Chapter Nine). So the key question becomes whether or not our reliable historical records are sufficient to give us a guide to the kind of extremes of weather and climate we can expect over the next fifty to a hundred years.

As a rule of thumb, we might guess that an understanding of the average weather of the past ten years could give a reasonable guide to the weather we might expect next year, that a century of recorded climate would be needed to enable us to guess the extremes of variation that might occur in the next decade, and that with a millennium of climatic history to interpret we might see the worst and best climates that might return within the next hundred years or so. That would be a conservative guide, in the sense that the longer the historical record we have the better the guide to variations of the present climatic systems will be, so that a thousand years or so is probably the very minimum period we need to have any guide to possible variations of the next century. Clearly, the more records we have the better – at least, up to a point. Once we look back to the past of 10,000 years ago we come to the end of the most recent Ice Age. Now, without doubt an Ice Age with extensive glaciers spreading across the regions of the Earth near the poles is something which doesn't fit into the pattern of climatic variations in recorded history. Something quite dramatic must change the climatic patterns into a new shape at the beginning of an Ice Age, and at its end. So we would not expect any information or records about climatic variations *during* the Ice Age to be much guide to climatic patterns *after* the Ice Age. An analogy might be made with a visitor from outer space studying the nature of a large city such as London or New York. Starting from the centre of the city, he would get a better and better idea of the average conditions in the city, and the extremes of variation within it, by looking at an ever-growing area – until, that is, he reached the limits of the city and began studying surrounding countryside. More and more information about farmland outside the geographical boundary of the city would be no use in finding out how much housing conditions vary inside the city; and more and more information about conditions before the historical barrier of the Ice Age will be of no use in finding out how much climatic conditions vary between Ice Ages.

So we have two boundaries on our historical records, if we want to guess the limits of climatic variations in the next century. We want at least a thousand years of records, plus as much beyond the thousand-year point as possible, up to ten thousand years ago. From then back, even if we have good guides to climate, they won't tell us much about what is likely to happen to us now – though they might, of course, tell us something about what might happen again within another ten thousand years or so. Historical records certainly cover more than our millennium-long minimum period, in one form or another, so we certainly should expect to learn something relevant to the climatic problems of the late twentieth and early twenty-first centuries from the lessons of history. Furthermore, various geological and geophysical techniques can be used to tell us something of how climate has varied right back to the latest Ice Age and beyond. These provide the basic guide for our expectations of just how much variation really is 'normal' in the perspective of recent centuries.

The long-term perspective

The longer-term records and their interpretation provide material for more than one complete book in themselves,* and it is not within the scope of the present book to go into details of just how these records are interpreted in climatic terms. They include, for

* See, for example, my own *Forecasts, Famines and Freezes.*

example, studies of how sea levels, indicated by erosion of sea-shores, have changed as the amount of water locked up in ice caps varies; the way sediments laid down in lake beds over the centuries change according to the amount of rainfall and flooding; varying thicknesses of tree rings affected by extremes of heat and cold, drought and flood as the trees were growing; studies of how whole species of plants have died out in a region as climate changes (indicated by the absence of their pollen from sediments deposited in lakes); and many more. The interpretation of the records is as much an art as a science – but when several different climatic indicators all point to the same broad conclusion about patterns of climatic change we can be sure that the conclusion is the right one. And it is this picture, painted with a broad brush and lacking in fine details, that we have for most of the past ten thousand years of climatic history.

Some tantalizing clues to perhaps regular patterns of variation emerge from this broad perspective. Variations in temperature which go up and down with periods of 2600, 1300, 650, 400 and 200 years have all been 'found' at one time or another, and if any of these could be proved as a reliable guide to temperature changes we would have a powerful forecasting tool indeed. But while some students of the changing climate can be found who accept the evidence for each of these cycles, there are no cycles that are accepted as proven by everyone who works on climatic problems. The main difficulty is that even if a small part of the pattern of temperature changes follows a regular cycle, the broader sweep of warming and cooling patterns clearly does not. The ripples on a pond may follow a regular pattern, but that doesn't tell you how deep the water is beneath. But the more cautious students of cyclic climatic variations perhaps deserve, at least a brief mention, and some recent work by Dr Bent Aaby of the Geological Survey of Denmark, leads to a conclusion broadly similar to that of many other students of cycles, and therefore to be taken seriously, even if it does not provide a precise guide to future prospects.

Dr Aaby's special study involved changes in the nature of the peat laid down in Danish bogs over the past 5500 years. Among other factors, changes of climate cause a shift from light to dark,

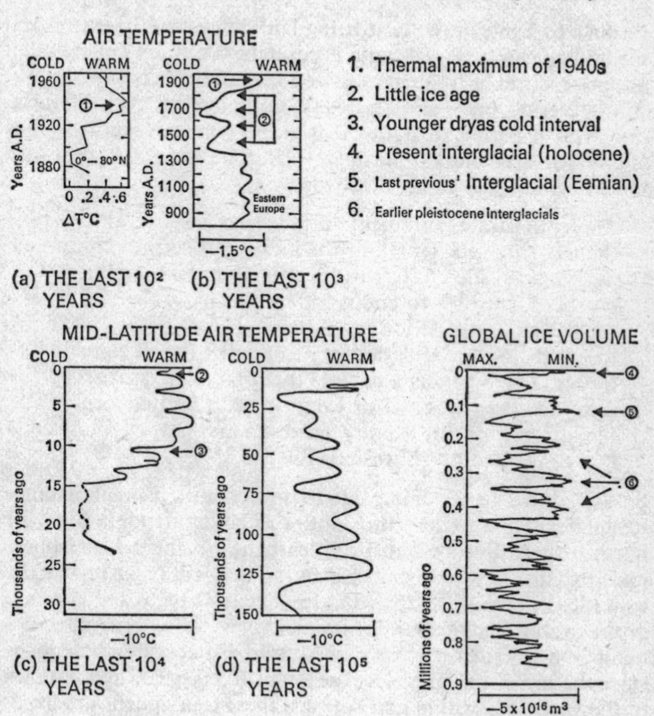

Figure 2.1 Different kinds of record show in overlapping segments how climate has varied over the past million years. The lesson is clear: climate is always changing, on a variety of different timescales. (Based on Figure A2 of the NAS publication 'Understanding Climatic Change' (see bibliography)).

Reproduced with permission of the US National Academy of Sciences.

or dark to light, in the peat being laid down, and layers of peat in the bog can be dated quite accurately. Dr Aaby believes that he has been able to single out the climatic effects on the peat from other changes, and he finds a fairly distinct cycle of variation with a period of about 260 years, which presumably has affected the whole of north-west Europe over the past five millennia, at least. In his own words:

> To demonstrate periodicity, a time span of generally more than 10 times the length of the individual periods concerned must be considered. In this investigation the time interval has been extended to about 5500 yr, or more than 20 times the length of the period in question. It is therefore justifiable to use the observed periodicity to predict future climatic trends. This suggests a general trend towards decreasing mean temperatures and/or wetter conditions may begin in the last part of this century or in the first part of the twenty-first century in north-western Europe.

So already we have some guide to how conditions might change in the near future – and this hint of a cooling trend, not just for north-western Europe but for at least the Northern Hemisphere and perhaps the whole globe, is one which will recur repeatedly as we study climatic history. But leaving aside the vexed question of the overall significance of climatic cycles, the records show unambiguously that there have been four distinct climatic epochs since the end of the latest Ice Age about 10,000 years ago. Studies of the variations within and between these four epochs are likely to give the best guide to extremes of climate in the next century.

The four periods are: the warm epoch immediately following the Ice Age, which peaked between 5000 and 7000 years ago; the colder epoch of the Iron Age, worst between 2300 and 2900 years ago; a little warming (sometimes called the Little Climatic Optimum) of the early Middle Ages, roughly at its peak between 1000 and 800 years ago; and a colder period since then (sometimes called the Little Ice Age), at its most severe from about 125 years to about 550 years ago, but perhaps not yet ended. More information is available for the more recent epochs, but the oldest of the four was the most different from present-day conditions, so is in some ways the most interesting.

From about 5000 to 3000 BC sea level rose rapidly as the last ice sheets of the former Ice Age melted, so that by 2000 BC the seas were 3 metres higher than today, implying that there were ten thousand billion* cubic metres less ice on the mountains then than there are today. The climate of the Sahara and deserts of the Near East was wetter than it is today, shown by a variety of records, and temperatures in Europe in summer were 2–3°C warmer than now. All this may have been linked with a northwards shift of the main high pressure (anticyclone) belt and the tracks of depressions (low pressure systems) moving around the North Pole – of which more later.

The next distinct epoch produced pronounced cooling especially between 900 and 300 BC, but even more striking is the increase in wetness (shown by peat bogs, for example) right across northern Europe from Ireland to Scandinavia. At this time, Russian forests spread southwards and summer temperatures fell. Further south, however, in the Mediterranean and North Africa the cooling was the main feature of the epoch, while rainfall may have been only slightly more than it is in that region today. Roman records showing that cultivation of the vine and olive was spreading northwards in the period after 100 BC seem to suggest strongly that in the immediately preceding centuries the weather had indeed been colder, and too severe for these crops in the north.

By the time of the Little Optimum, around AD 1000 to 1200, we begin to have much more comprehensive records, both historical and geological, geophysical and archaeological. Overall, this optimum followed the pattern of the warm period after the Ice Age, but it was shorter and less warm overall. In western and central Europe cultivation of the vine extended 3–5° further north in latitude, and 100–200 metres higher above sea level, consistent with average summer temperatures just over 1°C greater than today. Evidence from North America shows that the Little Optimum lasted there until about AD 1300, and further south there seems to have been a wetter period, corresponding to the European optimum, affecting Central America, Cambodia,

* billion = million million.

the Mediterranean and the Near East.

Finally, in our quick overview of past climates, we come to the most recent and potentially most disturbing epoch, the Little Ice Age. Generally regarded as the period from 1430 to 1850, but worst in Britain in the seventeenth century, this period was by and large one of harsher conditions than we are used to, separated by periods of more equable climate. The Arctic pack ice expanded considerably from its state in the Little Optimum, forests at high levels declined catastrophically, glaciers advanced in Europe, Asia Minor and North America, and there was generally a shift towards the equator of the rainfall belts (marked by the tracks of depressions) in the Northern Hemisphere – records from the Southern Hemisphere are sparse and difficult to interpret. These are all changes which could be catastrophic today, with our present food-producing capacity stretched by the ever-growing population of the world. And this is the cause for recent concern about climatic change among informed climatologists. Can we be sure that the relative ease of the past century marks the end of the Little Ice Age? Or could it just be a warmer spell which will soon end, returning us to conditions as bad for agriculture (let alone other effects on society) as those of the seventeenth century? As things stand now, we should be prepared for anything that has happened in the past thousand years or so to happen again, as far as climate is concerned, and the detailed picture of the past 1500 years of climatic change around the North Atlantic provides the guide we need to future fluctuations based on the lessons of the past. Before spelling these out, however, there is also something to be learned from events at the end of the warm period after the Ice Age, when the onset of the Iron Age cold epoch brought disastrous consequences to civilizations far removed from the polar regions where the cooling was worst.

The Indus Empire and the Iron Age cooling

The cooling, like other widespread coolings, went hand in hand with a shift in the circulation patterns of the atmosphere, bringing rainfall belts southwards towards the equator, and with a lid of cold air that sat over the pole growing and pushing southwards. We don't know why this so-called Arctic expansion occurred – we don't even know whether the expansion caused the cooling or was the result of it – but we do know the effects that it produced. Forests in Canada whose northern edges had reached closest to the pole in 1900 BC were pushed back by at least 200 km within the next century, giving us a very accurate date for the onset of the expansion. What this meant in terms of the overall circulation of the winds of the world we know very well from watching the atmosphere today – each year the pattern of winds around the Arctic (the circumpolar vortex) expands in winter and retreats in summer, giving us a view in miniature of how an Ice Age, or a Little Ice Age, would develop. With a bigger circumpolar vortex, the shift in rainfall belts pushes desert regions closer to the equator and the monsoon rains of Africa and India, the seasonal rains on which agriculture in those regions depends, cannot push so far inland. So the expectation of the climatologists is that any prolonged change in the circumpolar vortex will be important not just for the trees in Canada but even for farmers in the far-off lands of India and Africa. Professor Reid Bryson, of the University of Wisconsin-Madison, has related in detail the consequences of the Arctic expansion of 1900 BC for the great empire which then existed in north-west India, the Indus civilization based on the great Indus river but extending over a wide region of what is now the Rajputana Desert on the borderlands between India and Pakistan.

This was a great agricultural empire, with huge granaries which have been found in archaeological searches. The Empire existed in parallel with the Sumerians, with whom it traded, and was one of the most advanced cultures of the time. Yet, in the period following 1900 BC the cities were abandoned and taken over by drifting sand dunes, returning to desert for a period of seven centuries. The monsoons had failed, and the sand moved in over the agricultural land. Freshwater lakes turned first to salt, then dried up completely. The whole might of the Indus Empire fell before a shift in the winds, associated with the expansion of the circumpolar vortex. And a very similar pattern of change in the monsoon winds also occurred in North Africa, even though there was no comparable civilization there to be brought to its knees as a result. Here is another lesson from climatic history, albeit rather a long time ago by human standards. Clearly, as Professor Bryson points out, it is quite possible, within the overall pattern of climatic changes since the latest Ice Age, for the monsoon rains to fail not just for a year or two, or for a decade, but for a full seven centuries. And it is with this reminder that there could be even worse climatic events than those revealed by studies of the past few centuries that we turn now to the detailed historical and climatic records of the past 1500 years.

Recent climates and cultures around the North Atlantic

For the past millennium and a half, the information climatologists have gleaned about events around the North Atlantic, and the evidence from history and archaeology, provides us with a detailed picture of the way small changes in climate (by the standards of the past 10,000 years, let alone compared with the very long term) can have profound effects on human society. A team of Danish researchers headed by Professor W. Dansgaard has provided us with one record of climatic changes over this period that is not only detailed but complete, indicating temperature variations from year to year in Greenland over the complete span of the past 1420 years.

This remarkable record comes from analysis of ice samples drilled from the Greenland glacier; a core of ice 404 metres long includes layers built up each year by the successive winter snows, and each winter snow carries within it a record of the prevailing temperature at the time, in the form of different kinds of oxygen atoms. The oxygen we breathe comes in two forms, identical except for the weight of their atoms, called oxygen-16 and oxygen-18. Where oxygen is combined with hydrogen to form water, this weight difference in the oxygen atoms means that there is also a weight difference between two families of oxygen molecules – and at a given temperature the rate at which the two families evaporate differs by a precise amount which determines the proportion of the two remaining to be locked up in water, ice or snow.

This is the principle underlying the use of ice cores as long-term 'thermometers'. It is no mean trick to pick out from the ice core a slice corresponding to a particular year's snowfall, to melt the ice and analyse the water produced in terms of the proportions of the two oxygen atoms (isotopes) it contains. But once it is done, the result is a very accurate temperature guide indeed, whose accuracy can be checked and calibrated against the ordinary temperature records of recent decades, which show that it is indeed reliable. So, without worrying too much about the details of the technique (except to tip our proverbial hats to the people who did the work) we can use this temperature record in our survey of the lessons of climatic history, looking in particular at the changes in Norse society occurring in the region of the North Atlantic close to the site of the Greenland ice core 'thermometer'.

There is, of course, a fair amount of historical material, especially from the Sagas, which tells us about changes in the Norse culture just over a thousand years ago. One interesting result which is clear from the broad picture this produces is that both the Little Optimum and the Little Ice Age seem to have been at their peaks in Greenland a couple of hundred years before their counterparts in America and Europe. The reasons for this are obscure, but if the pattern continues then European and American climatic conditions for the coming decades may follow the patterns from Greenland a couple of centuries ago . . . which would mean a general cooling trend over the next hundred years. This means little in itself, but does fit in with some other indications of how the present climate is changing. The details of the effect of these changes on the Norse are, however, even more striking.

According to the Landnam Saga the first attempt to settle in Iceland was made in AD 865 by a farmer, Floke Vilgerdson, just at the time we can now see as the end of a short series of cold fluctuations. He lost his cattle in a severe winter, and came home with tales of 'a fjord filled up by sea ice . . . therefore he called the country Iceland'. But within another ten years successful settlers had reached Iceland, flourishing in the growing warmth until in 985 they were sending their own explorers and would-be colonists westwards. According to the Greenlander Saga, Eirik the Red, the founder of the Greenland colony, gave Greenland its name in a deliberate attempt to encourage settlers – what we would now call a confidence trick. But the Greenlander Saga was written down hundreds of years after Eirik's time, when the climate had become much colder. What the ice core thermometer tells us is that Eirik in fact reached Greenland near the end of a warm period, longer than any warm period that has occurred there since. So, in all probability, the coastal regions at least really were green, and his naming of the land was apt. The historical accident that Iceland and Greenland were discovered in that order, at a time of climatic warming, is probably the reason why they have the names we know them by, when anyone who has visited them both must wonder why the names are not the other way around!

During the last phases of the Little Optimum, Norse explorers and perhaps even settlers reached North America. But this was the limit of the westward expansion of Norse society. Historians have argued about the exact causes of the collapse of this widespread culture, but the onset of the Little Ice Age must have been, at the very least, a major contribution. More sea ice as the weather deteriorated, the glaciers pushing south, and frosts keeping more and more land frozen all the year round spelled the end of the American adventure, the Greenland colony, and very nearly the Iceland colony as well. They also brought the end of the historical records from these areas, which then still depended on the verbal tradition of the Sagas, with no written records for historians to puzzle over today. So we shall never know just what happened in the last days of the Greenland colony. But we can see how this shift into Little Ice Age conditions affected the climate of North America, and changed the culture of at least one group of American Indians, at the same time that it brought the end of the great days of Norse exploration.

Once again, it is Professor Bryson we have to thank for gathering the threads of the story together. The broad feature of the change in climatic terms was a cooling of the Arctic after

AD 1200, with the sea ice pushing south and an expansion of the circumpolar vortex, reminiscent of both the earlier, bigger shift that brought disaster to the Indus region and the smaller shift that is worrying us today. In the North American continent these changes moved the summer westerlies southwards and brought a reduction in the rainfall of the northern plains and the present-day corn belt. All this is shown by various climatic indicators, Professor Bryson's special study being the analysis of the changing vegetation of a region indicated by the changing pollen remains left in different layers of sediments. And the effects on farmers living in the region at the time were disastrous.

The Mill Creek Indians were one of many groups of farming people established on the plains on the eastern side of the Rocky Mountains where at present spring wheat provides a staple crop. Their villages were established, so the archaeological record shows, around AD 900, a time when the climate for farming in the region was improving. By AD 1200 this was a relatively rich society, the people living in a region of tall-grassed prairie with wooded valleys, growing corn and hunting the deer that abounded in the region. But in the span of only 20 years all this changed. The rains disappeared as part of the Arctic expansion; the tall grass died out and was replaced by short grass; the deer could no longer find food and died out in the region; the farmers' crops failed; and the people were left to depend on bison for their food – and indeed for their whole way of life. The settled villages of wealthy farmers were replaced by a hunting society of poorer people, moving to keep within reach of their only reliable source of food, the moving herds of bison. This shift in world climate, then, was the reason why these Indians reverted to the way of life that we now regard as 'typical' of American Indians before the arrival of European settlers, and produced the myth perpetuated in so many films and TV westerns.

But the big lesson here is not that climate can and does change. We are now beginning to realize that change is the normal state of climatic affairs. What Professor Bryson draws attention to as the lesson of this climatic change is that it brought 200 years of drought to the main grain-growing regions of North America. This region is regarded as the 'breadbasket' of North America, and its produce is today essential not just for Americans but for a desperately large proportion of the world's population that has come in recent years to depend on exports of American grain to live on. The lesson, in Bryson's words, is that 'clearly two hundred years of drought in the "breadbasket" of North America is possible'.

And such changes are not isolated geographically. At the same time that these disasters affected the Norse settlers and American Indians, Europe suffered from cool, damp conditions that brought widespread outbreaks of blight, destroying crops while, again as Bryson points out, the heavy, clay soils of the English midlands became too wet to work effectively and were largely abandoned by the early part of the fourteenth century. As far as we can tell, the overall population of the world (at least, the then-civilized world) declined during this climatic deterioration. The historical records are still poor, as far as relating climatic and cultural changes goes, at that time. But the dramatic increase in information that is available for each passing century means that we can stay with the changing European picture to find out the best details of what happened during the worst ravages of the Little Ice Age.

The most dramatic examples of the extreme cold come from England in the seventeenth century. This was the time of the great English diarists, and John Evelyn, in particular, has left us almost a day-by-day account of the weather which is well worth reading (the edition edited by E. S. de Beer is recommended). This and other sources tell us about the great Frost Fairs, when the river Thames froze solid, the greatest freeze of all occurring in the winter of 1683–4. Evelyn's diary for the late seventeenth century contains many references along the lines of 'great drought' (June 1681), 'intolerable severe frost' (1683–4), 'excessive cold' (November 1684) and 'backward spring' (1688). The climatologist Professor Gordon Manley has gathered anecdotes from a variety of historical sources which indicate just how bad things were by the standards we have become used to, and some of these are well worth repeating here.

In January 1698, John Locke, the philosopher, recorded a temperature of 38°F in his parlour at about 10 a.m.; on 12

January 1684, ink froze in the bottle at the fireside of an Oxford scholar. But as Professor Manley points out the real test of a severe winter in those times came when travellers, writing home about their experiences at wayside inns on the road, recorded that certain 'vessels' were found frozen under the bed in the morning! And what of droughts, which have made so much news lately? Gilbert White of Selbourne, writing in his diary on 29 September 1781, noted that 'my well has now only three feet of water; it has never been so low, not since my father sank it, more than forty years before'. And that remark came just *before* a very dry October which produced just half an inch of rainfall, on average, over England and Wales.

The beginning of instrumental records of temperature, pressure and rainfall, means that for the period since the middle of the eighteenth century, at least, it is possible to work out just what was happening to the atmosphere and its circulation to produce such extremes. With a little guesswork, the picture can be drawn a little more sketchily for earlier periods as well. What we find from this analysis of wind flow patterns is that the underlying cause of such extremes in the late seventeenth century and on to about 1712 was a frequent occurrence of so-called 'blocking' situations, in which a sluggishly moving atmospheric circulation gets stuck into one pattern, with persistent high pressure bringing, in this case, cold northerly winds over the region of Iceland, Scandinavia and the British Isles. The high pressure belt in some of the most severe winters of the Little Ice Age extended from Iceland or north-north-eastern Europe to Spain or the Azores, a distribution that has not been seen on any monthly average pressure map since that time. The next stage of this work, in progress at the Climatic Research Unit of the University of East Anglia, will extend this understanding of detailed weather patterns back into the warmth of the Middle Ages, using material from monastic records, ships' logs, state papers and so on. But already we can begin to see from the historical evidence how the patterns of the atmospheric circulation can shift, bringing changes in rainfall and temperature as a consequence. And here is another lesson from history. The changes in circulation that go hand in hand with a cooling of the

Arctic region bring a shift into the more sluggish pattern of circulation, the 'weak' circulation. And because this weak circulation pattern allows the weather machine to get stuck in one gear for long periods of time, it brings with it many extremes of both temperature and rainfall.

This is seen by the example of the worst winters of the Little Ice Age in Britain. The cooling of the whole Northern Hemisphere which was associated with the expansion of the circumpolar vortex and onset of weak circulation conditions was not on its own enough to bring very severe winters. But because the weak circulation could get stuck in a pattern of northerly winds over Britain, say, some of the winters were very severe indeed by British standards. Not only had the Arctic cooled a little bit, but in effect the Arctic winds were often allowed to blow unchecked much further south than they could otherwise. Of course, the weak circulation pattern can also get stuck for months at a time in a different gear, bringing perhaps not a vicious winter but a very dry summer to Britain, with the high pressure systems blocking out the wet Atlantic air for months. This is exactly what has happened in the summers of 1975 and 1976, bringing hot, dry summers to England and Wales at a time when many climatologists believe the Northern Hemisphere is cooling down. What we learn from such paradoxical events is that a slight cooling down is not in itself very important for patterns of climate, except in a few regions, one of which is, ironically, the key grain-growing region of the world today. Much more important, in most parts of the globe, is the change to weak circulation and erratic variations of weather from season to season and year to year. Nowhere is this better highlighted than in the events in England and Wales in 1976, when the worst drought since records began was followed by the wettest October/November for more than seventy years. The prospect of droughts followed by floods is enough to make any farmer despair – and such a prospect is much more likely when the circulation of the atmosphere weakens as the circumpolar vortex expands.

The reasons why such changes occur will be discussed later, but in surveying the more recent history of climate some details become clear which should be kept in mind. There was a relatively cool period at the beginning of the nineteenth century, just a decade or two, which is recent enough for good records to be available. Professor Lamb has calculated that the change in climate at that time can best be explained by a reduction in the amount of heat reaching the surface of the Earth, by about one or two per cent, together with shifts in circulation of the kind outlined above. So history also tells us that we need to find some way of explaining how the heat reaching the Earth's surface can change by up to two per cent in the space of a few years.

The abnormality of the twentieth century

Within the perspective of the past millennium, we can now see that the global climate of the twentieth century has been most unusual. What happened first was the reverse of the shift outlined above, with a return towards stronger circulation, more steady, predictable weather, a retreat of Arctic ice and contraction of the circumpolar vortex, and an overall warming of the Northern Hemisphere. This brought good conditions for agriculture and food production at a time of exploding population, helping to boost that explosion. It also, unfortunately, meant that our ideas of what is 'normal' in climatic terms have been coloured by the weather of the first half of the twentieth century – a half century which Professor Lamb describes as being the most abnormal of the past thousand years.

In the early 1950s, indeed, there was concern lest the warming trend that had just been identified should continue indefinitely, with the ice caps melting and raising sea level to flood cities such as London and New York. Many studies point to the dramatic size of this brief warming, and singling out one in particular we can look at changes in the altitude at which trees have been able to grow in Norway. Dr J. A. Matthews, of the University of Edinburgh, has interpreted this evidence in a reconstruction of changing summer temperatures since 1700, since, of course, trees grow at higher altitudes when the climate is warmer. Dr Matthews's calculated temperatures, relative to the period 1949–63, are shown in Table 2.1 for each 25-year period since

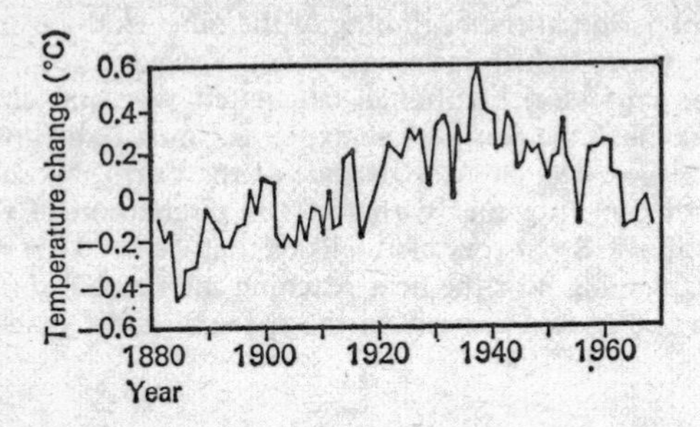

Figure 2.2 Recorded changes of annual mean temperature of the northern hemisphere over the past 90 years, originally determined by M. I. Budyko and brought up to date by H. Asakura.

Reproduced with permission of the US National Academy of Science.

1700; in his words 'the twentieth century (1901–50) is indicated as being warmer than any previous 50- or 25-year period, with summer temperatures that were on average only 0·3°C below those of 1949–63'.

That would be all very well if we could take such evidence as indicating that the Little Ice Age is over, a new climatic pattern is established, and we can expect the warmth to continue for decades to come. Unfortunately, that isn't so, as the events described in Chapter One have already made clear. Since the early 1960s, we have seen only too clearly the shift towards weaker circulation, erratic extreme local variations of weather, and a slight cooling of the Northern Hemisphere that we now know signals a return towards the expanded circumpolar vortex conditions of the Little Ice Age. The analogy you draw depends on where you live: Professor Lamb has pointed out that the climatic change we are now experiencing is a shift towards the conditions of the seventeenth century in Britain, and away from the abnormal conditions that have persisted through most of the twentieth century; Professor Bryson, from his base in Wisconsin, prefers to point out that the present shift is in the same direction as the climatic change which brought two hundred years of

Table 2.1 Differences between summer temperatures for each 25-year period from 1700 to 1950. The difference is measured in degrees C compared with the average for the period 1949–63 and indicates just how much the conditions we think of as 'normal' differ from the conditions that have prevailed in recent centuries.

Date (AD)	Index (°C)
1701–1725	−1·63
1726–1750	−1·14
1751–1775	−0·48
1776–1800	−0·87
1801–1825	−1·15
1826–1850	−0·46
1851–1875	−0·55
1876–1900	−0·43
1901–1925	−0·28
1926–1950	−0·34

Taken from J. A. Matthews, *Nature*, vol. 264, p. 243 (1976).

drought to the world's breadbasket less than a thousand years ago. Neither prospect is exactly comforting. There is no evidence to suggest that the recent warm peak was anything other than an unusual hiccup, a brief alleviation of what may be a continuing Little Ice Age. Paraphrasing Bryson, the lessons of climatic history can be spelled out as follows:

CLIMATE IS NOT FIXED

CLIMATE TENDS TO CHANGE RAPIDLY RATHER THAN GRADUALLY, as the Norse, Indus and Mill Creek people found to their cost

CHANGES IN SOCIETY ACCOMPANY CHANGES IN CLIMATE

WHEN HIGH LATITUDES COOL WE HAVE MORE ERRATIC EXTREMES OF CLIMATE

OUR IDEA OF 'NORMAL' CLIMATE IS MOST ABNORMAL IN TERMS OF THE PAST MILLENNIUM, and we are now seeing a shift back towards the cooler, more erratic pattern

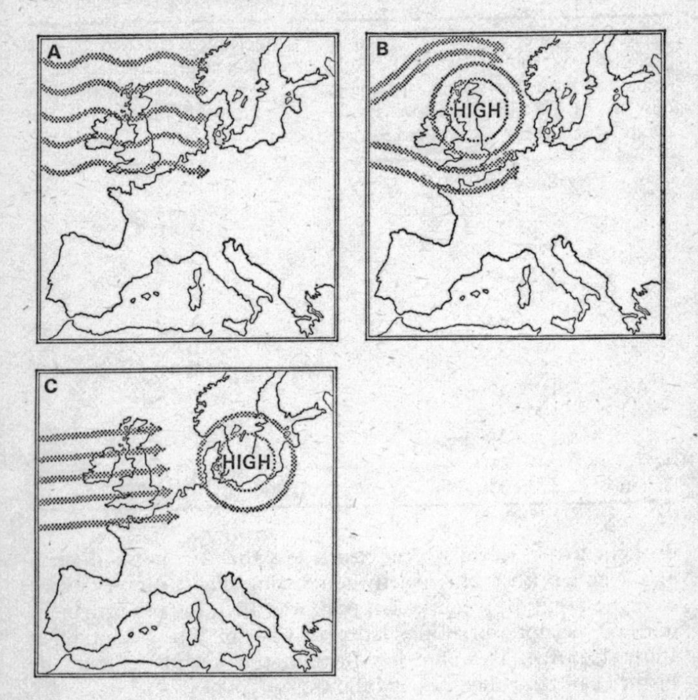

Figure 2.3 Outline map of UK and Europe to show blocking effects. (a) When the atmospheric circulation is strong, depressions follow regular paths across the Atlantic, depositing moderate rainfall over the UK and moving on. (b) With a blocking anticyclone over Britain, depressions are diverted. This explains why, for example, Britain had drought in 1976 while Baltic regions had unusually heavy rain. (c) With a blocking system to the east of Britain, depressions run into a barrier over the UK and stop, depositing excessive rainfall and causing floods. So the occurrence of droughts followed by floods in the UK is an example of the effects of a weak circulation with common blocking systems.

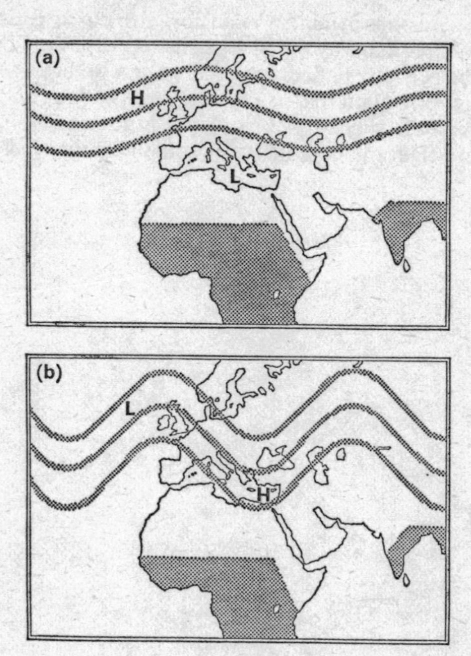

Figure 2.4 Diagram indicating principal differences between extreme circulation types and associated rainfall patterns. (a) Strong zonal circulation produces high rainfall over Britain and allows monsoon rains to extend well to the north in India and Africa; (b) weak zonal circulation leads to low rainfall over Britain and suppresses the northward extent of the monsoon rains. (H=High rainfall; L=low rainfall; the change in the boundary of the dotted area shows the change in the extent of the monsoon region) (Dr D. Winstanley, *Nature*).

With these lessons in mind, we can now turn our attention to the problems of how and why the climate changes – for only by understanding the processes of climatic change can we hope to predict the likely future patterns of climate, and find out if, as the historical records suggest, we really are in for something much less pleasant than the bulk of the twentieth century has been so far.

Chapter Three

Why does the climate change?

We all have a vague idea of climate as some kind of 'average weather' that is typical of a particular place on the surface of the Earth. Now, with the evidence of history before us, we can also see that this average weather depends on just where in time – where in history – we are observing the climate, as well as just where we are geographically. And as well as local climatic variations we are aware that the whole pattern of weather and climate across the globe can change in world-wide patterns of interlocking climatic variations on all scales up to and including the switch from Ice Age conditions to non-Ice Age conditions. But why does the climate change? If we could answer that one in satisfactory detail, then we need never be surprised by any future changes of climate; unfortunately, there seem to be many different processes affecting climate world-wide, and the trick is going to be not only finding them but also picking out the one or two effects important for the fifty years we are most interested in – that is, the next fifty years! Still, with the upsurge of interest in, and study of, climatic change that has occurred in the 1970s (for obvious reasons) we can make a better stab than ever before at unravelling the complexities of climatic change.

I do not intend here to go into any detail about the workings of the atmosphere of our planet, which are covered in many standard texts including some mentioned in the bibliography of this book. Basically, the atmosphere can be considered as a heat engine, or weather machine, driven by the heat energy the Earth receives from the Sun plus a winding-up effect from the rotation of planet Earth. The small amount of heat that leaks out to the surface from the centre of the Earth just doesn't matter compared with the energy input from the Sun. This heating is strongest at the tropics, near the equator, where the Sun is more nearly overhead for much of the year, and weakest near the poles, where the Sun hangs low on the horizon. So the tropics are hotter than polar regions, and air circulates from equator to poles and back in a continuous attempt to even things up. The pattern is complicated by the fact that land heats up more effectively than the oceans (and cools off quicker in winter when the Sun is less effective) and that different kinds of land (forest, mountains, farmland or deserts, for example) absorb heat differently. Finally, the changing amount of snow and ice cover at high latitudes affects heat absorbed dramatically, since a shiny white snowfield or ice cover reflects both heat and light away into space. And as if that were not enough, clouds and dust in the atmosphere, and changes in the atmosphere itself, can all affect the amount of heat getting through to the Earth's surface. Fig. 3.1 sums up all the interacting components that combine to produce the climatic system; it's small wonder that whole textbooks are needed for a proper explanation of even the present-day situation. But since our interest is in the changing face of the climatic system we can for simplicity leave aside the details of how these interactions are interpreted and concentrate instead on the effects produced by modifying individual components of the complex system.

Ocean water and polar ice

One of the most pressing needs before a complete understanding of climatic change becomes possible is an understanding of the circulation of the oceans, and of the way atmosphere and oceans interact. After all, most of the Earth's surface is covered by water, and like the atmosphere the oceans respond to the differential heating of the Sun at different latitudes by developing currents which have the overall effect of transporting heat from regions near the equator to cooler regions at high latitudes. The ocean currents have their own 'weather' systems, curling masses of water equivalent to the high and low pressure regions of the atmosphere, but the overall role of the oceans in determining climate is only just beginning to be understood. Certainly, we

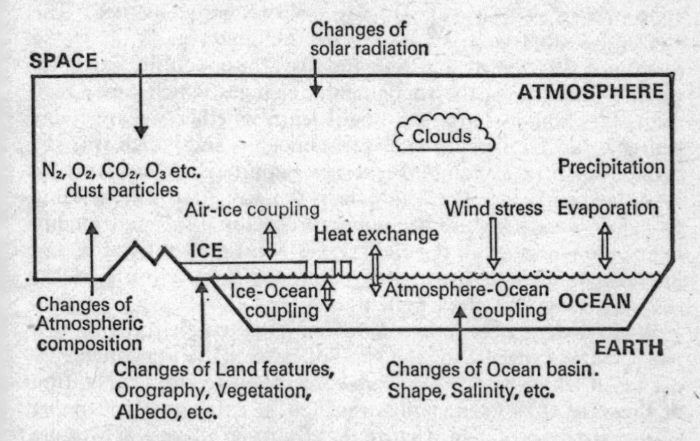

Figure 3.1 Schematic illustration of the components of the coupled atmosphere–ocean–ice–earth climatic system. The full arrows are examples of external processes, and the open arrows are examples of internal processes in climatic change.

Based on Figure 3.1 of NAS book 'Understanding Climatic Change'. Reproduced with permission of the US National Academy of Sciences.

cannot yet interpret the fine tuning of the ocean's contributions to climate – the sort of effects that might be important in the next fifty years. But we can now begin to understand ways in which oceanic processes could play a part in the most dramatic climatic change of all, the onset of a full Ice Age. It seems worthwhile to look at the details of this interaction, as they are best understood today, both to provide a picture of the importance of the oceans to climate and as an example of the strengths, and limitations, of computer modelling techniques.

There are two ways to start looking at what happens to the weather and climate of the Earth as time goes by. You might try to work out what the weather is going to be like in your home town tomorrow, on the basis of what the weather was like today and yesterday. This is the point from which all conventional weather forecasts start, and it is an extremely valuable, practical approach where our day-to-day routine is concerned. The alternative approach is to start by thinking big. What is the difference between an Ice Age and a warm epoch on Earth? If we can track down the fundamental changes which cause such dramatic climatic shifts, we might learn whether we are today nearer to an Ice Age or to a warm epoch – and which way the climate is shifting. Combining the two approaches, and working from each end towards the middle, is the way meteorologists and climatologists today are beginning to develop a full understanding of the workings of the climatic system. From both ends, and in the middle, it is increasingly clear just how big a part the oceans play in the whole system.

With modern electronic computers, it is possible to build up a mathematical 'model' of the state of part of the atmosphere or ocean, in which sets of numbers correspond to the temperature or pressure at different points through, say, the atmosphere, or to the strength of ocean currents at chosen places. If we had enough of these numbers, spaced closely together, we would have in effect a numerical model of the real system, and by adding the known laws of physics to the model it can be 'evolved' from a chosen starting point to predict future states of the system. This requires, however, a lot of numbers for points very close together in order to work effectively – and the ocean–atmosphere system is very big. A model to describe the development of weather systems needs a very powerful computer, which can operate at very high speeds, otherwise we find that it takes the computer more than 24 hours, perhaps, to evolve the weather system through an amount of evolution corresponding to one day in the real world! So computers can only be a help to us in understanding climatic change and weather variations in two extreme cases. The smaller the weather system being studied, the more accurate the model will be, and computers are indeed now being used as a great aid to straightforward weather forecasting for local regions of the globe ('local' here being anything up to a whole state or country, but smaller than a continent). At the other extreme, we can apply models to the whole of the Earth, or to one hemisphere, deliberately leaving out a lot of detail and trying only to get a very rough idea of the way the climate changes if we disturb the system in some way. Such general circulation models are good enough to reproduce broad features of the real atmosphere, such as the monsoon and trade winds, or the pattern of prevailing westerlies sweeping depressions across the North Atlantic. So they can give a rough guide to, for example, the amount of rainfall produced around the world under different conditions. This has recently solved one of the biggest controversies in climatology for many years – were Ice Ages wetter or drier than the present day? One theory had it that more precipitation (rain plus snow) would be necessary in an Ice Age, to build up the glaciers; the rival view essentially held that with water locked up in ice and a cooler planet overall there would be less evaporation and so less precipitation. Either theory sounds plausible, but when a general circulation computer model corresponding to the present day is compared with an equivalent model in which the ice cover of high latitudes is set to match the pattern of the most recent glaciation the results show clearly that there was less rainfall in the Ice Age.

So, in climatic studies, computer models are very useful at answering specific, general questions – as long as we have wit enough to ask a question which is important and simple, that will produce a clear-cut answer. The computers are also, of course, very valuable in taking away the drudgery of repeated numerical calculations in any kind of mathematical work, not just in detailed model building. But we always must bear in mind that a computer that is a million times faster than a man will just make a million times as many mistakes as a man if it is fed with the wrong numbers somewhere along the line.

Getting back to the link between ocean currents and Ice Ages, we need to bring in not just ocean–atmosphere interactions, but ocean–ice interactions. The idea is that when the ice sheet over Antarctica around the South Pole builds up beyond a critical size it will break up, and that this break-up will affect the circulation of ocean waters in a way which brings on Ice Age conditions. And the model offers a neat explanation of one puzzle which has emerged from the traces left by past Ice Ages – why should the tropical waters of the Atlantic be *warmer* in an Ice Age than they are today?

At present, these Atlantic waters are cooled by upwelling deep water that has moved out from the Antarctic along the ocean floor, starting in a spinning current, or gyre, in the Weddell Sea, just at the southern edge of the Atlantic. When the ice cover of the West Antarctic region breaks up, the currents flow quite differently, so that the tropical Atlantic is no longer cooled in this way – and at the same time the spread of snow and ice cover on land further north may be encouraged.

But if the Antarctic ice broke up tomorrow, we wouldn't feel the full effects immediately. It takes the cool ocean floor current a thousand years to complete its journey to the warm, tropical Atlantic, and while the last of the cool water was on its journey increased evaporation from the sea uncovered by the broken West Antarctic ice sheet would lead to increased precipitation, rebuilding the ice sheet, or beginning its rebuilding, before the effects of the break-up were felt in the north. Eventually, the warmer water of the Atlantic, no longer cooled by the deep Antarctic current, would interact with the atmosphere to change the whole pattern of atmospheric circulation, encouraging, according to the models, the spread of snow and ice in the north. Once this spread begins, it can hurry the world into an Ice Age, because land or sea covered by new snow or ice reflects away more solar heat than before, so that cooling proceeds in a vicious circle, or positive feedback process.

At first sight, it seems difficult to get out of this feedback loop. But eventually the ice sheets spread out beyond the land surfaces and over the seas, and at the same time the Antarctic ice is building up again, re-establishing the old ocean water circulation from the Weddell gyre to the tropical Atlantic. When the northern ice sheets get too big, they break up – and just at a time when the tropical Atlantic is again being cooled by Antarctic water. So things revert to what we think of today as 'normal'.

It's a nice idea, and it ties in many pieces of the climatic interaction in one model. Most probably, however, even this model doesn't go far enough in that direction, and we still cannot see just what decides the moment for the disintegration of the Antarctic ice sheet. Perhaps this model explains the way some Ice Ages start, or perhaps we need something else to give the ice a push at the right time. The best overall picture of how outside influences might bring the onset, and end, of Ice Ages seems to me to be that the amount of heat reaching the surface of the Earth at different latitudes varies in some way, and this will be discussed fully later (Chapter Seven). But any theory that suggests Ice Ages start when the Southern Hemisphere cools off, for example, would still need some explanation along the above lines for what happens to the ocean, ice and atmosphere as the Ice Age develops. Already, though, this model has hinted at something of crucial importance in our discussion of climatic change. For the theory only works if we have a continent at the South Pole, surrounded by ocean, and a lot of land at high latitudes in the Northern Hemisphere, where snow can build up into new glaciers. One of the most important developments in our understanding of planet Earth in recent years has been the realization that this geography has not been the same for all time – that the continents drift about the face of the globe. So before we look specifically at how changes in the atmosphere affect climate, we should take a brief look at the much longer-term perspective, the way present-day patterns of climatic change depend on the present-day geography of our planet.

Continental drift and climate

The amount of heat reaching the tropics is quite enough to keep the whole of our planet free from ice – provided warm water from the tropical oceans can circulate near to the polar regions and keep them ice free. But in the south today the pole is surrounded by land, which is therefore covered in ice; and the North Pole lies almost in the middle of a shallow sea, which is itself almost surrounded by land. Warm currents just cannot penetrate into the Arctic sea, so there too a layer of ice has formed. Either polar continents or land-locked polar seas seem to be essential for our watery planet to have any ice caps at all, and even there it seems there are complications involved.

As we have already seen, the high reflectivity of ice and snow – their high 'albedo' – can play a big part in maintaining an Ice

Age, by reflection of the Sun's heat out into space. And according to some calculations so much heat is reflected away from the Arctic regions today that without the ice cover the Arctic sea would be as much as 40°C warmer than it is at present – warm enough, in fact, to stop the ice from forming at all. Strictly speaking, our present-day climate is that of a relatively warm period *within* an Ice Age, an Ice Age that is being maintained, in a very depleted form, by the very presence of the ice cover at high northern latitudes. This has some intriguing consequences, not least being the very real possibility that if we could find some way of blackening the Arctic ice (perhaps by dusting it with soot?) we could remove it once and for all, since blackened ice would absorb solar heat and melt. Such an idea might not go down too well in Europe or the US, where a rise in sea level of perhaps 10 metres, produced by melting the ice, would inundate densely populated cities and valuable farmland; but it could be better than the spread of ice back into a full Ice Age, and on a timescale of thousands of years such a possibility could turn out to be a real alternative for our global society. Fortunately, however, such problems are not important when we are focusing our attention on the next half century or so.

Only twenty years ago, it would have seemed ridiculous to discuss the implications of a different distribution of the continents for the climate here on Earth. Until the dramatic change in our understanding of the Earth brought about by new discoveries and new theories in the 1960s, the established view was that continental masses did not shift about on the surface of the globe, but could only move vertically, raising up mountains or sinking to form seas but not shifting sideways. On that picture, it might seem rather remarkable that the Earth should have one pole surrounded by land, and the other with a polar continent sitting across the pole, but at least the theories held that this had always been the case. That being so, we could accept this status quo as the background to the climatic story, and look elsewhere for reasons why some epochs should be prone to Ice Ages while others – for hundreds of millions of years at a time – were completely ice free.

But now all that has been changed. In its modern form, under the new name of 'plate tectonics', the idea of continental drift has become established as fact. It is now established beyond reasonable doubt that continents do move sideways across the surface of the globe, carried by the movement of large blocks of the Earth's crust ('plates') which include both continents and the material of the ocean floors. Given this new evidence, the climatic situation has a new degree of complexity. For most of the Earth's history, there simply were no land-locked polar oceans, and no polar continents. The warm ocean water from the tropics could sweep up to both poles unimpeded, and the spread of warmth could keep the planet ice free. That, in a nutshell, is why Ice Ages are rare features in the history of our planet. Indeed, today we can use evidence of glaciation on old rock layers to interpret the drift of continents in the past, completing a feedback loop of scientific understanding!

Acceptance of the idea of continental drift as a reality for planet Earth dates back only to the mid-1960s, but one of the key features of geology which the concept explains is the varying evidence of glaciation in different parts of the world. With hindsight, it now seems obvious that traces of glaciation in rocks now found in tropical regions mean that those rocks have drifted through polar regions, not that ice once covered the entire globe. With a more detailed understanding of the history of continental drift, geophysicists have found that extensive glaciation does indeed occur only when a continent lies across a pole, or when a polar region is surrounded by continents. The present situation, which has persisted for about two million years, is unusual in that *both* poles have conditions suitable for the formation of ice sheets, at the same time. So, just as we found in Chapter Two that the twentieth century is hardly typical, in climatic terms, of the past thousand years, so we find from geophysics that the past two million years is an unusual climatic epoch compared with the whole history of our planet, since the time when atmosphere and oceans first formed. Today, average temperatures at middle latitudes are perhaps 10°C more than in the depths of the most recent Ice Age – but still 8 or 9°C cooler than is normal in terms of the history of the Earth over the past hundreds of millions of years.

Calculations of the changing albedo of the Earth as continents drift from low to high latitudes have been performed at Leicester University, and these show another contribution which also acts to encourage Ice Ages when the continents are near the poles. As well as the important differences caused by preventing warm ocean water from penetrating to high latitudes, it turns out that there is a strong cooling influence when the continents move away from the equator, because they are no longer able to absorb so much of the Sun's heat. With oceans dominating the tropics, more heat is reflected away into space, because water has a greater albedo than land.

Quite apart from its interest in the long term to geophysicists, astronomers and others concerned more with abstract theories than with the practicalities of present-day life, all this tells us something of key importance to understanding climatic variations today. Throughout most of the history of the Earth, it seems that the distribution of the continents simply was not right for ice to form. Any small changes in the amount of heat getting through to the ground could be shrugged aside – whatever caused them. But in the very unusual state that the world has been in for the past couple of million years, a small reduction in temperature, for whatever reason, can allow ice to form and spread. And once this happens, the reflective power of the ice can help things to cool down even more. So we can see that even small changes in the heat arriving at the surface of the Earth are significant in the present epoch. It is no wonder, on this picture, that climatic variations of all kinds have been common for the past two million years, and in particular it is no surprise to find, as we did in Chapter Two, a continual record of changing climate throughout historical time. The presence of polar ice is the key to climatic change, and the positions of the continents are the cause of the existence of polar ice. Given this situation, small-scale 'trigger' effects are sufficient to affect climate dramatically. So, what are the possible triggers – how can the amount of heat reaching the surface of the Earth vary?

The radiation balance of the Earth

No one is quite sure just how much effect on temperature is produced by a small change in heat arriving at the Earth's surface – the computer models are not good enough to give a precise answer. Various calculations suggest that a 1% change in heat put into the Earth's climatic system might change average temperatures by anything from 0·6 to 1·2°C; so as a rule of thumb we won't be too far off if we bear in mind that a 1% change in heat input is roughly equivalent to a 1°C change in temperature. That is a worrying situation, since it means that a change of only a few per cent in terms of heat input would be enough to bring on a full Ice Age, given the peculiar present-day arrangement of the continents, which gives us a North Pole surrounded by land, and a South Pole covered by a continent. It's even more worrying when we find out just how many processes there are which affect the heat balance of the Earth, and although only a couple of these will be found to be important enough to consider in detail the others are there as a reminder that there are more checks and balances involved in the climatic system than we can yet interpret fully.

First, of course, there is the radiation (heat) coming in from the Sun. Traditionally, astronomers have regarded our Sun as a stable star, burning steadily at the same heat all the time – at least, on any timescale important for human affairs. Unfortunately for this complacent view, however, we have no direct evidence that this is the case, since measurements of the Sun's output are only accurate to within about 2%, leaving ample margin of error for heat changes which could explain, for example, the Little Ice Age and the optimum of the mid-twentieth century. Recently, astrophysics has been turned on its head by a flurry of discoveries that the Sun is not as simple as astrophysicists used to think, and the whole question of solar variability is now raised as a respectable puzzle for astronomers. This puzzle will be discussed in Chapter Five. Even if the radiation reaching the top of our atmosphere from the Sun is constant (and my own interpretation of the evidence is that it is not) the amount of heat absorbed by the Earth can vary according to circumstances. Some of the radiation is absorbed high in the atmosphere by gases which warm up and then radiate heat away again into space – perhaps 6% of incoming radiation departs in this way immediately. Then, about 20% of the incident radiation is reflected away by clouds, while about 16% of the heat is absorbed by the water, dust, ozone and other constituents of the atmosphere and 3% is absorbed in warming up clouds. Finally, about 4% of the heat is reflected away immediately by the ground, leaving just half (51%) of the radiation that arrived at the top of the atmosphere to be absorbed by the oceans and continents of the Earth's surface.

Even this heat must be returned out into space, of course, since otherwise the Earth would warm up through absorbing more and more heat energy. Indeed, what has happened is that the present temperature of the Earth, overall, has adjusted automatically, like a global thermostat, until the heat being radiated is in balance with the heat being absorbed. This radiation occurs as direct long-wave radiation, through heat being carried up into the atmosphere by convection and then radiated, and by changes involving the heat given up by water vapour as it condenses or freezes, and the heat absorbed by melting snow or evaporating water. Anything which happens in the atmosphere to change the amount of heat penetrating it (changes in cloud cover, in the number or nature of the particles in the atmosphere, its chemistry, and so on) will change the heat balance of the Earth, so that the surface warms or cools until a new balance is reached. And, of course, changes in the surface of the Earth itself (melting ice or freezing snow, clearing forests to make farmland, damming rivers to produce shiny lakes, and so on) will have a similar effect on the overall heat balance. In the light of all this, the wonder is not that the climate varies so much, but rather that it shows any signs of stability at all!

So it is no wonder that climatologists do not attempt to include all these factors in their models at once. Instead, one or two of the variable factors, or parameters, are adjusted in the models while the others are kept steady, and the way the overall picture changes is taken as a guide to how much the real radiation balance depends on those selected parameters. There must always be a nagging doubt that some unforeseen interaction between the parameters makes the results of this kind of modelling invalid, but it's the best we can do. There is also another nagging doubt, that should in honesty be mentioned here, before we begin to look at some of the results of the detailed modelling.

It has been argued that all the variations in climate that we observe in the historical record are the result of 'random noise' in the climatic system. By random noise in this connection, statisticians mean any and all of the fluctuations which occur simply because there are so many particles (atoms and molecules, dust, snow and so on) interacting with one another in the climatic system. Just by chance, each interaction between a pair of particles is going to produce one result from an array of possible results, in a much more complicated version of the 'experiment' that involves tossing a coin a thousand times or more. If the same truly balanced coin is tossed repeatedly, it will produce a succession of 'heads' and 'tails', a so-called time-series in which although on average we find half the tosses give heads and half tails, there may be runs of tens or scores of tosses in which either heads or tails dominate. If the time-series is plotted out as a graph, it can produce a dramatic pattern of variations – even regular-seeming 'cycles' of heads, followed by tails, followed by heads again – although we know that the record is that of a simple random choice experiment. The number of such either/or 'choices' occurring throughout the atmosphere every second is immense, and this must produce a random 'noise' which affects our observations of the varying climate. Such noise must provide the ultimate limit on how well we can interpret past climatic changes and forecast future climatic changes, so that our forecasts can never be 'perfect'. But it doesn't really seem very likely that *all* climatic fluctuations are produced by random noise, since the modellers have recently achieved some considerable successes in explaining some of the patterns of past climatic changes in terms of some of the parameters involved in the radiation balance of the climatic system. If the models work in explaining past climates, they should be just as good (but not any better) at predicting future climates; so there lies the best hope for an understanding of the nature of the climatic threat we have to face in the next fifty years or so. And the reason why we can pull out a coherent explanation from the tangle of different parameters is very much that the different parameters don't all act at once. When we are looking at the millions of years important to geophysics, we need not worry too much about small variations in the way the atmosphere transmits heat to the Earth below; when we are looking at a timescale of thousands of years we don't need to worry about continental drift; and when we are looking at prospects for the next half century we don't even need to be concerned too much with, say, changes in the ice cover of Antarctica. The answers we need in the immediate future can only come from looking at the parameters in the climatic system which directly affect the amount of heat getting through to the Earth's surface from year to year – changes in the atmosphere and changes in the Sun itself.

Chapter Four

Changes in the atmospheric sunshield

Of all the varying parameters which can change the amount of solar heat getting through to the ground or sea, regardless of any changes in how much actually arrives at the top of the atmosphere, two in particular now seem to be of vital importance for changes over a few years or a few decades. One of these, the influence of dust from volcanoes acting as a sunshield in the air, has been discussed by climatologists for as long as they have been aware of climatic changes. The other, the effect of changes in the ozone layer of the stratosphere, has only been appreciated within the past couple of years as a serious contender in the climatic stakes. But both depend very much on the overall structure of the atmosphere of the Earth, a layered structure in which the air nearest the ground, the troposphere, and the layer just above, the stratosphere, play the key roles in establishing patterns of weather and climate.

This layered structure can be seen best by looking at how the temperature of the atmosphere varies at different heights above the surface of the Earth (Fig. 4.1). We have already seen in passing that solar heat gets bounced back into space and absorbed and re-radiated in the atmosphere, with only just over half actually getting through to the ground. But although the atmosphere keeps some of the solar heat from ever reaching the ground, it also acts as a blanket to keep the Earth warm, making efficient use of the heat which does get through. The warm ground, or sea, heats up the air just above by radiation at heat wavelengths, in the infra-red. This infra-red radiation is strongly absorbed by carbon dioxide and water vapour in the air, and the warm lower atmosphere radiates heat in its turn in all directions, some going upwards and eventually out into space, but some going back

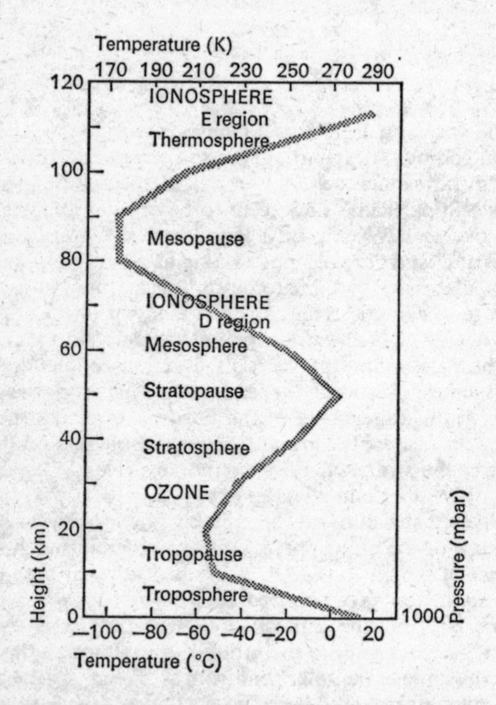

Figure 4.1 Variation of atmospheric temperature with height showing layering of the atmosphere.

down to the surface and helping to keep it warmer than it would be with no blanket of air above. This is the so-called 'greenhouse effect', which definitely operates in the atmosphere, although a minor controversy has been stirred up lately by claims from some scientists that this is not the way greenhouses themselves keep warm.

The warmth of the ground is also responsible for convection in the lower atmosphere, since the heated air expands, becoming less dense than the air above, and rises up through convection. This convection is a key feature of the processes that produce weather and climate, contributing to the overall circulation of the atmosphere. But the warm air cannot rise upwards to the top of the atmosphere, because it is held down within the lowest layer, the troposphere, by the presence of the stratosphere above. The stratosphere has been described as a 'lid' on the troposphere, holding down convection and keeping weather confined within the lowest layer of the atmosphere. Clearly this lid is as important for the overall patterns of weather and climate as the occurrence of convection is – and it operates because the stratosphere is an 'inversion layer' in which convection does not occur.

Right through the troposphere temperature falls with increasing height by about 6°C for every kilometre of altitude gained. The fall slows down at about 10 km and stops altogether at 15 km. From 20 km altitude to about 50 km temperature then *increases* from −60°C at the top of the troposphere (the bottom of the stratosphere) to a maximum of about 0°C at the top of the stratosphere. Such a warming can only mean that the air in the stratosphere is gaining heat energy from somewhere, and that energy must be coming from the Sun. The effect of this inversion – the 'inverse' of the normal pattern because instead of decreasing temperature increases with increasing height – is to stop convection. The old adage that 'hot air rises' is true only if the air above the hot air is cool; if the hot air lies under a layer of even hotter air, then it cannot rise and there is no convection. Of course, we have to adjust our usual thinking about just what 'hot' means here; in fact, the air at the bottom of the stratosphere, with a temperature of −60°C, is held down by the presence of 'hot' air above with temperatures in the range from

−60°C all the way up to the temperature of freezing water!

This increase in temperature occurs because solar heat is constantly being absorbed in the stratosphere in a series of reactions involving the production and break-up of ozone molecules. Ozone is the molecule of oxygen with three atoms (O_3) instead of the more familiar two (O_2), and it has reached public awareness lately through various scare stories about the harmful effects for life that might result if all or some of the ozone in the stratosphere disappeared through man's inadvertent interference with natural processes.

The first scare about mankind's activities disrupting the ozone layer came with the advent of high-flying supersonic aircraft, which actually fly through the stratosphere. The fear was that a large fleet of aircraft like Concorde could disturb the natural balance of the ozone layer, by throwing out large quantities of polluting nitrogen oxides, gases which combine easily with ozone and can break it down into ordinary oxygen. This scare soon passed – accurate calculations soon showed that hundreds of such aircraft, in daily use, would be needed to cause any long-lasting damage to the ozone layer, and as we all know only too well there just aren't that many Concordes around to worry about!

But another, more disturbing prospect appeared in the middle 1970s. This is the threat from spray cans – so-called 'aerosol' sprays. All such sprays contain two ingredients, the active material which we buy the cans for (shaving soap, deodorant, paint or whatever) and a gas which does the pushing to get the active ingredient out when we press the button. As of 1976, about half of all aerosol sprays used compounds called fluorocarbons (FCs) as propellants, to do the pushing, because the FCs are very stable, non-inflammable and non-toxic. Unfortunately, it now seems that with enough FC aerosols around so much of these compounds will be released that they will get into the stratosphere in significant quantities. And, you've guessed, one of the few reactions the FCs do participate in fairly eagerly is the break-up of ozone.

For once, though, a very real threat seems to have been spotted in time. Several US states, led by the pioneering legislature in Oregon, have now banned the use of FCs as propellants in aerosol sprays, and with ever-growing public awareness of the hazard the aerosol industry is rapidly changing over to other propellants even where not yet required to do so by law. Yet again, however, as one scare passes another looms up. Now, the doomsdayers argue that overuse of nitrogen fertilizers on agricultural land is producing a build-up of nitrogen oxides in the atmosphere which could do the same damage as a hypothetical fleet of hundreds of Concordes. But that does look a bit like a red herring – after all, the importance of the supposed threat from aircraft was that the oxides were released right in the stratosphere itself. Ground-level disturbances, while they might filter up to produce some effect, are unlikely to completely disrupt the natural balance. The point is that while the ozone layer is very important to life on the Earth's surface, natural fluctuations are much bigger than anything yet likely to result from mankind's interference, and life has adapted over millions of years to cope with changes on the scale of those natural fluctuations.

With no ozone shield, ultra-violet light from the Sun would penetrate through to the surface of the Earth, with damaging consequences for life, burning up crops and causing skin cancers in people and other animals. With no ozone, there would also be a dramatic change in the nature of the lid on top of our weather systems, with the prospect of drastic changes in the circulation of the atmosphere. All in all, we clearly need to keep the ozone where it is, since it is doing a good job. But in fact there is very little prospect that anything mankind can now achieve in terms of affecting ozone in the stratosphere can compete with the natural changes that are taking place all the time through the input of solar energy.

Oxygen molecules in the stratosphere absorb solar ultra-violet radiation to produce free oxygen atoms (O) as the paired atoms in O_2 are broken up by the energy. Then, each atom may combine with an oxygen molecule (O_2) to build up an ozone molecule (O_3). All this depends on there being enough oxygen around, and enough ultra-violet energy. So a balance is struck intermediate between the surface, where there is most oxygen, and the top of the atmosphere, where there is the strongest flux of solar energy. All the time more ozone is being built up in this way – but we don't have an atmosphere full of ozone because all the time ozone is reacting with other substances, giving up its extra oxygen atom and reverting back to the diatomic molecular form, O_2. The details of this dynamic process are complex, and the way in which ozone concentrations in the atmosphere might change through outside influences is not at all obvious. What is obvious is that this dynamic balance is held roughly in balance by a series of natural processes, and that the nature of the ozone layer – and of the whole atmosphere – today is the result of adjustments and evolution that have occurred over hundreds, even thousands, of millions of years. The existence of an ozone layer is like the existence of a river; the water in the river is constantly changing as the rains fall and water runs into the sea. But the river itself remains a reality, with changes in level from time to time. In just the same way the ozone layer has existed as long as there has been oxygen in the air, with new ozone molecules always being added and others taken away, and with fluctuations in concentration of ozone analogous to the changing level of a flowing river.

The time taken for equilibrium balance to be reached depends very much on the exact altitude of different parts of the layer. Above 50 km, it takes only a few minutes to reach equilibrium because there is very little gas at all to be balanced. Below about 30 km, where the air is denser and more molecules are involved in the balancing, it can take several days for a new equilibrium to be reached after something changes the old balance. So the lower region of the stratosphere, always disturbed by the top of the troposphere and the atmospheric circulation below, never really reaches a true equilibrium, while at higher altitudes there is a noticeable shift in the balance even from day to night – but not in the direction you might expect.

The surprising observation that ozone concentration through the whole depth of the atmosphere seems to increase at night emphasizes the dangers of drawing 'obvious' conclusions about how changing outside influences will affect the ozone layer. What seems to happen is that above 40 km the air is so thin that very little 'new' ozone is being made, while a great deal of 'old' ozone is being destroyed by the intense solar radiation. The high layers of the stratosphere are kept supplied with ozone only because molecules migrate upwards from deeper levels. So, at night, the balance of the equilibrium is shifted *in favour* of ozone, while at lower altitudes although the balance is shifted against ozone there is no time for the shift to take effect before the night is over. The deep layers show the average influence over many days, or around the world, while the upper layers reflect the present situation more accurately. Already it should be clear that changes in the output of solar energy must affect the ozone layer, and therefore have an effect on climate and life more complicated than the direct influence we might expect. In working out overall climatic influences, it seems sensible to start near the ground, with volcanoes, and then to investigate the upper atmosphere and solar effects, so details of effects will be held over for now. But before we return to the ground to look at the parameters which seem most significant for predicting climatic changes of the next half century or so, it seems a shame not to mention, if only in passing, how the layers of the atmosphere above the stratosphere blend into space.

From 50 km to 80 km altitude, another cooling layer dominates the now rapidly thinning atmosphere, with the coldest atmospheric temperatures reached at the top of this layer, the mesosphere, at about −100°C. From there on outwards temperature increases in the last thermal layer, another inversion layer called the thermosphere. Again, in this region energy is absorbed from the Sun, but now the solar energy is so intense that the process goes beyond the simple dissociation of oxygen molecules that occurs in the stratosphere. At the top of the atmosphere, not only are molecules broken into atoms by solar energy, but atoms themselves are partly ionized, with negatively charged electrons being knocked off from them to leave behind positively charged ions. For this reason, a more useful way to describe the upper atmosphere is in terms of the extent to which atoms have been ionized, and the whole region is therefore dubbed the ionosphere, with sub-layers corresponding to different degrees of ionization.

Where charged particles begin to dominate the atmosphere, the Earth's magnetic field becomes more important than the feeble pull of gravity at such altitudes, and the region of space near to the Earth where these effects dominate is the magnetosphere – the last boundary of anything that might be considered as the region of our home in space. But space beyond the magnetosphere is far from empty, as space probes have shown. All the time there is a flow of charged particles – electrons and ions – more or less outwards from the Sun. This solar wind may also play its part in affecting the workings of the atmosphere, and more of this in the next chapter. But now, having reached the boundary of 'Spaceship Earth' we can begin again from the ground upwards, looking not at the structure of the atmosphere but at the parameters working within the atmosphere to affect climate by modifying the strength of the sunshield between us and the arriving solar heat energy.

Dust in the atmosphere

A layer of dust which acts as a sunshield in the atmosphere does more than just block out sunlight and heat from reaching the ground. The dust itself warms up as it absorbs some of the solar heat, making the layer of atmosphere where the dust is thickest warmer than it would otherwise be, while the ground cools off below. Indeed, it's not absolutely certain that the ground will always cool, since the warm dust itself radiates heat away, and some of this will reach the ground – but at least some of the warmth from the dust layer must be radiated off into space, so there should be some cooling below, even if not as much as might be expected if the 'sunshield' just took away a certain fraction of the heat arriving from the Sun. Just how big an effect dust in the atmosphere can have on climate depends both on the nature of the dust and on where in the atmosphere it is. Dust from blowing soil, pollution by man's activities, smoke from forest fires and particles thrown out in volcanic eruptions are all among the candidates put forward to explain climatic changes of one kind or another, and although we shall look at some of the human influences in Chapter Nine, when it comes to natural climatic variations caused by dust only volcanoes seem able to explain some features of the historical record of climate in terms of changes in the Earth's natural sunshield. As much as anything else, this is because large volcanic eruptions are able to put a lot of dust up into the stratosphere, where it can produce a relatively long-lived effect. Dust which is blown into the troposphere soon gets washed out of the air and back down to ground by rain; but dust in the stratosphere, above the weather layer of the atmosphere, can persist for much longer. Professor Lamb, who has made a detailed study of the links between historical volcanic eruptions and climatic changes, divides eruptions that throw dust into the stratosphere into two categories: in the first, dust reaches the lower stratosphere, about 20–27 km altitude; in the second, dust is blown much higher, perhaps to 50 km altitude. The first group is regarded as the most important for climate, because very great quantities of material can reach the lower stratosphere, producing dense, long-lived veils of dust; the second category is of interest to specialist climatologists, since it may help to explain formation of very high ultra-cirrus clouds, at altitudes as great as 80 km, but seems to be of little direct importance for the broad features of the changing climate, simply because very little dust ever gets so high.

The effects of the dust, carried round the lower stratosphere by high altitude winds which sweep across the whole globe, can be very large. The measured intensity of the Sun's heat* fell by some 20–30% after great eruptions in 1883 (Krakatoa), 1902 and 1912, and according to Professor Lamb an estimated reduction in surface temperature of 1·3°C followed the occurrence of two great volcanic eruptions in 1783. Perhaps significantly, the middle part of the twentieth century, which we have already noted as the warmest run of decades for centuries, occurred at a time when there were very few volcanic eruptions, and so very

* When we are dealing with changes in the dust in the air, rather than changes in the heat leaving the Sun itself, the relationship with temperature changes becomes a little more complicated. While there may be much less measured heat *at ground level* from the direct rays of the Sun, most of this 'lost' heat is actually being absorbed by the dust in the air, and only a little has been reflected away into space to be lost altogether. So the heat is redistributed, as well as diminished, and the fall in temperature overall is much less than simple measurements of direct solar intensity might frighten us into expecting.

little dust around to shield us from the warmth of the Sun. But looking at individual eruptions, and particular features of the changing climate, doesn't really give us much of a clue about what is happening. The broad relationship between volcanoes and climate can only be seen by looking at the whole pattern of volcanic activity for hundreds of years – and this is just what Professor Lamb has done.

Before any progress could be made in this direction, someone had to establish a comparison between the veiling effects of the dust from different eruptions; Lamb's Dust Veil Index, which defines the veiling effect of the 1883 Krakatoa eruption as 1000 and measures other eruptions relative to this standard, now provides this basic comparison for the past 300 years, and establishes a standard against which even older eruptions can be approximately compared. Lamb finds that there is a small but significant effect of major eruptions on the atmospheric circulation. This is in line with records that suggest that many of the coldest, wettest summers in Britain, North America and Japan occur in 'volcanic dust years'; Lamb says that 'it may be that volcanic dust has played a part in all the very worst summers of these centuries [seventeenth to twentieth]' and that 'it seems clear that there is also a tendency for cold winters in Europe after some volcanic eruptions, evidently those in low latitudes, that produce world-wide dust veils'. The key word here, however, is 'tendency'. For no one is claiming that the changing DVI is alone responsible for all the climatic changes that have occurred in the past three or four centuries. Lamb's work, and other studies, show that volcanic dust contributes to climatic change, but, again in Lamb's own words, 'nevertheless volcanic dust is not the only, and probably not the main, cause of climatic variations within the period surveyed'. We must look for other effects to combine with, or oppose, the dust-veil effects in order to explain the historical record of climate. But before we begin to look for such other effects, and even though we are mainly concerned with effects that operate over a few tens of years, we cannot leave the story of dust in the atmosphere without at least a mention of one very dramatic prospect – that volcanic dust may play a part in the development of full Ice Ages.

Table 4.1 Different sizes of dust particle stay in the stratosphere for different times – and the effect of dust in the atmosphere also depends on how high the particles are thrown by volcanic eruptions. The figures given here, taken from H. H. Lamb's book *Climate: Present, Past and Future*, indicate how long it takes dust to fall down to the tropopause (a) for tropopause at 17 km altitude and (b) when the tropopause is at 12 km altitude. 1 micron (μm) is one millionth of a metre.

Particle diameter	*Initial height (km)*	*(a)*	*(b)*
2 μm	40	25 weeks	41 weeks
	30	21 ,,	37 ,,
	25	16 ,,	31 ,,
	20	7 ,,	23 ,,
1 μm	40	1·9 years	3·1 years
	30	1·6 ,,	2·8 ,,
	25	1·3 ,,	2·4 ,,
	20	0·6 ,,	1·7 ,,
0·5 μm	40	7·8 years	12·5 years
	30	6·5 ,,	11·3 ,,
	25	5·0 ,,	9·7 ,,
	20	2·2 ,,	6·9 ,,

Although this idea seems appealingly simple, it is, to say the least, controversial. Professor Bryson, for example, has said that 'the evidence . . . is quite convincing', while Dr B. J. Mason, Director-General of the UK Meteorological Office, believes that the result of a computer modelling experiment in which a layer of dust put into the model atmosphere succeeded in warming the model stratosphere without producing any discernible effects in the model at ground level 'hardly supports the thesis that cooler epochs in the historical record may have been caused by volcanic eruptions'. Dr Mason points also to the fact that a volcanic

Table 4.2 Lamb's Dust Veil Index assessments and chronology of major volcanic eruptions cited

Year	*Volcano*	*Situation*	DVI
1680	Krakatoa	6°S 105½°E	400
1680	Tongkoko, Celebes	1½°N 125°E	1000
1693	Hekla, Iceland	64°N 19½°W	100
1693	Serua, Molucca Is.	6½°S 130°E	500
1694	Amboina, Molucca Is.	4°S 128°E	≮250
1694	'Celebes'	1°N–6°S 119–125°E	≮250
1694	Gunung Api, Molucca Is.	4½°S 130°E	400

NOTE: If the low temperatures prevailing in England, as well as Iceland and a wide surrounding region, over the years 1694–8 were representative of a world-wide anomaly of about the same amount, and provided their departure from the temperatures prevailing in the immediately preceding and following years were entirely due to volcanic dust, the total DVI for 1694–8 should be 3000–3500.

Year	*Volcano*	*Situation*	DVI
1707	Vesuvius	41°N 14°E	150
1707	Santorin	36½°N 25½°E	250
1707	Fujiyama, Japan	35°N 139°E	350
1712	Miyakeyama, Japan	34°N 139½°E	200
1717	Vesuvius	41°N 14°E	100
1717	Kirishima Yama, Japan	32°N 131°E	200
1721	Katla, Iceland	63½°N 19°W	250
1730	Roung, Java	8°S 114°E	300
1744	Cotopaxi, Ecuador	1°S 78°W	300
1752	Little Sunda Is., possibly Tambora	8°S 118°E	1000
1754	Taal, Luzon, Philippines	14°N 121°E	300
1755	Katla, Iceland	63½°N 19°W	400
1759	Jorullo, Mexico	19°N 102°W	300
1760	Makjan, Moluccas	½°N 127½°E	250
1763	'Molucca Is.'	2°N–3°S 125–131°E	600 (?)
1766	Hekla, Iceland	64°N 19½°W	200
1766	Mayon, Luzon, Philippines	13½°N 123½°E	2300 (?)
1768	Cotopaxi, Ecuador	1°S 78°W	900
1772	Gunung Papandayan, Java	7½°S 108°E	250
1775	Pacaya, Guatemala	14°N 91°W	1000 (?)

Table 4.2 – *continued*

Year	*Volcano*	*Situation*	DVI
1779	Sakurashima, Japan	31½°N 131°E	450
1783	Eldeyjar, off Iceland	63½°N 23°W	
	Laki and Skaptar Jökull, Iceland	64°N 18°W	700
1783	Asama, Japan	36½°N 138½°E	300
	Total veil, 1783:		1000
1786	Pavlov, Alaska	55½°N 162°W	150
1795	Pogrumnoy, Umanak Is., Aleutians	55°N 165°W	300
1796	Bogoslov, Aleutians	54°N 168°W	100
1799	Fuego, Guatemala	14½°N 91°W	600
1803	Cotopaxi, Ecuador	1°S 78°W	1100 (?)
1807–10	Various, including Gunung Merapi, Java	7½°N 110½°E	(?)
	and São Jorge, Azores	38½°N 28½°W	(?)
	Total veil, 1807–10:		1500 (?)
1811	Sabrina, Azores	38°N 25°W	200
1812	Soufrière, St Vincent	13½°N 61°W	300
1812	Awu, Great Sangihe, Celebes	3½°N 125½°E	300
1813	Vesuvius	41°N 14°E	100
1814	Mayon, Luzon	13½°N 123½°E	300
1815	Tambora, Sumbawa	8°S 118°E	3000
	Total veil, 1811–18:		4400
1821	Eyjafjallajökull, Iceland	63½°N 19½°W	100
1822	Galunggung, Java	7°S 108°E	500
1826	Kelud, Java	8°S 112½°E	300
1831	Giulia or Graham's Island	37°N 12–13°E	200
1831	Pichincha, Ecuador	0°S 78½°W	(?)
1831	Babuyan, Philippine Is.	19°N 122°E	300
1831	Barbados	13°N 60°W	(?)
	Total veil, 1831–3:		about 1000
1835	Coseguina, Nicaragua	13°N 87½°W	4000
1845	Hekla, Iceland	64°N 19½°W	250
1846	Armagora, S. Pacific	18°S 174°W	1000
1852	Gunung Api, Banda, Moluccas	4½°S 130°E	200

Table 4.2 – *continued*

Year	*Volcano*	*Situation*	DVI
1856	Cotopaxi, Ecuador	1°S 78°W	700
1861	Makjan, Moluccas	$\frac{1}{2}$°N 127$\frac{1}{2}$°E	800
1875	Askja, Iceland	65°N 17°W	300
1878	Ghaie, New Ireland, Bismarck Archipelago	4°S 152°E	possibly 1250
1883	Krakatoa	6°S 105$\frac{1}{2}$°E	1000
1888	Bandai San, Japan	38°N 140°E	250
1888	Ritter Is., Bismarck Archipelago	5$\frac{1}{2}$°S 148°E	250
	Total veil, 1883–90:		about 1500
1902	Mont Pelée, Martinique	15°N 61°W	100
1902	Soufrière, St Vincent	13$\frac{1}{2}$°N 61°W	300
1902	Santa Maria, Guatemala	14$\frac{1}{2}$°N 92°W	600
	Total veil, 1902:		about 1000
1907	Shtyubelya Sopka Ksudatch, Kamchatka	52°N 157$\frac{1}{2}$°E	150
1912	Katmai, Alaska	58°N 155°W	150
1963	Mt Agung (Gunung Agung), Bali	8$\frac{1}{2}$°S 115$\frac{1}{2}$°E	800
1966	Awu, Great Sangihe, Celebes	3$\frac{1}{2}$°N 125$\frac{1}{2}$°E	150–200
1968	Fernandina, Galapagos	$\frac{1}{2}$°S 92°W	50–100
	Total veil, 1963–8:		about 1100
1970	Deception Is.	63°S 60$\frac{1}{2}$°W	(200)

From *Climate: Present, Past and Future.*

eruption in Bali in 1963 produced dust in the stratosphere sufficient to cause a rise in temperature of several degrees, again with no detectable effect on temperature at ground level. Bryson and Mason perhaps represent the extreme views on the nature of the influence of volcanic dust on climate. Somewhere in the middle ground, various studies have now shown that layers of dust occur in ice cores drilled from glaciers, at just the depths corresponding to cold periods in the history of the Earth, and similar ash layers found in cores drilled from the sea bed have been interpreted as further evidence that some periods of increased glaciation, at least, coincide with times of increased volcanic activity.

Perhaps here, though, the key word is 'coincide' – for there is nothing to prove that the volcanoes *caused* an increase in glaciation. Indeed, one rather neat explanation of the coincidence is that when ice sheets grow and spread over the northern continents the mass of ice pressing down in the Earth's crust triggers off regions of volcanic instability, producing a lot of volcanoes and a lot of dust and ash as an *effect* of the glaciation! Perhaps we should be indeed thankful that the problem of the Ice Ages is not really our concern for the immediate future (but see Chapter Seven); the overall situation regarding volcanoes and climate, including the long-term perspective, was ably summed up by Drs Dragoslav Ninkovich and William Donn, writing in the journal *Science* in November 1976:

> There seems little doubt that many great eruptions close in time would charge the stratosphere with sufficient *aerosols* to seriously obscure total solar radiation, thereby causing a prolonged decrease in global temperature. So far, there is little in the historical record to indicate that such frequent events have occurred. The little ice age (~1500 to 1850) is the most significant temperature change documented since the climatic optimum, but there is no ready correlation between this event and prolonged and frequent great eruptions, which averaged a few per century in historic times.
>
> It may well be that the immediate consequences of global temperature decreases following explosive eruptions are both local and temporary in nature. For example, recent droughts in both Africa and the United States have been ascribed by some to global cooling related to solar variations. There is no reason why they should not also have been brought on by cooling from dust veils.

Solar variations plus the effects of dust veils? This is a very promising idea which we shall look at further in the next chapter. After all, no theory of how the climate is changing at present can be regarded as satisfactory if it fails to explain the Little Ice Age, the most significant climatic change of the past millennium. But first, we have still to consider ways in which changes in the high atmosphere, the stratospheric lid on top of the troposphere, can affect the strength of the solar heat penetrating to the ground, even if the amount arriving at the top of the atmosphere is constant.

Ozone and the stratosphere – a variable filter

Because it filters out the ultra-violet radiation from the Sun, the ozone layer makes life as we know it possible on the surface of the Earth. With no ozone in the stratosphere, many species would be wiped out, and only sea creatures would be safe from harm. So, quite apart from its climatic importance, the ozone layer has long-term and widespread significance for life on Earth. Whether or not man's activities are likely to damage the layer in the next few decades (and the best evidence now seems to be that this is rather unlikely), there may be natural variations in the concentration of ozone which affect living things and the climate. The most extreme possibility is that the layer might be wiped out altogether, and although this is again something which seems unlikely to happen within the next fifty years or so there is evidence that it has happened in the past – and that such an event may even have contributed to the disappearance of the dinosaurs from the face of the Earth some 65 million years ago.

This idea depends on the discovery that the Earth's magnetic field completely reverses its direction from time to time, with North and South magnetic poles swopping over. During this process, there is a period of perhaps thousands of years in which there is little or no magnetic field – when the Earth is no longer protected from the charged particles of the solar wind by the shielding effect of the magnetosphere. The abrupt way in which whole groups of fossils disappear from the geological record at certain times in pre-history has long been a major puzzle for palaeontologists, and in the 1960s it was discovered that out of eight such 'mass extinctions' of prehistorical species six had occurred at the same time as geomagnetic reversals. Could this be a coincidence? It seemed unlikely, but it was not until 1976 that a really plausible explanation appeared – an explanation which depends on variations in the ozone content of the stratosphere.

Dr G. C. Reid and colleagues working in Boulder, Colorado, have pointed out that the removal of the shielding magnetosphere could lead to the breakdown of ozone, which in turn would allow damaging ultra-violet radiation to penetrate to the ground until the magnetic field built up again (in the opposite direction) and ozone could again build up in the stratosphere. When particles from the solar wind penetrate into the Earth's atmosphere, they will collide with the atoms and ions there, producing a shower of energetic particles. One of the effects of this shower of particles will be to encourage the formation of nitric oxide at high altitudes – and nitric oxide has a singularly good ability to remove ozone through a pair of chemical reactions which leave the nitric oxide itself unchanged and ready to break down more ozone. First, a molecule of nitric oxide (NO) combines with ozone (O_3) to produce nitrogen dioxide (NO_2) plus ordinary molecular oxygen (O_2). Then, the nitrogen dioxide (NO_2) can interact with any free atom of oxygen (O) to produce nitric oxide (NO) plus another diatomic oxygen molecule (O_2).

The particles from the Sun which could do most damage in this way occur in bursts associated with flares on the surface of the Sun. A typical flare would be sufficient to account for enough ozone to allow 15% more ultra-violet than at present to reach the surface of the Earth, provided there was no shielding magnetic field to keep the particles at bay.

A superflare 100 times bigger than this, occurring during a geomagnetic reversal, could remove so much ozone that the ultraviolet radiation at the surface of the Earth would be more than doubled – and such a superflare is quite likely to occur once or twice during the hundreds or thousands of years when the Earth has no effective magnetic shield.

The biological effects could be great, since it seems that many small organisms now live precariously close to the limits of ultra-violet radiation that they can tolerate. And also, of course, such a dramatic change in the stratosphere would inevitably affect the overall circulation of the atmosphere, producing a climatic change for the creatures and plants to contend with just at a time when they are also being bombarded by extra ultra-violet radiation. The accumulation of effects could rapidly build up into the last straw which proves too much for many species and produces the observed extinctions in the fossil record – with as a bonus a convenient explanation that not all geomagnetic reversals coincide with such extinctions because sometimes the Earth gets through the reversal at a time when the Sun is not producing any superflares.

There seems to be a lesson here, too, which may be relevant to the problems we are likely to face in the next half century. For, although we do have a protecting magnetosphere at present, and most of the effects of particles in the solar wind are kept out of the atmosphere, some of this solar radiation does get through, especially at the time of large solar flaring activity. Without going so far as to strip off all our ozone and wipe out whole species, these more ordinary changes may be modifying the ozone filter enough to account for some observed climatic effects.

Solar flares are more common in some years than in others, and the whole pattern of activity on the surface of the Sun and in the solar wind which blows across space varies in a roughly regular way, over the so-called solar cycle which is roughly 11 years long on average, but can be anything from 9 to 13 years long (or even shorter or longer on rare occasions). At the beginning of a cycle, the Sun is quiet, showing few sunspots and little in the way of flares; then the activity builds up, reaching a maximum half a cycle later before dropping off into the next minimum, some 11 years after the beginning of the cycle. And just as more particles from the solar wind penetrate into the Earth's atmosphere when the shielding magnetic field is weak, so more particles get through the magnetic field when the solar

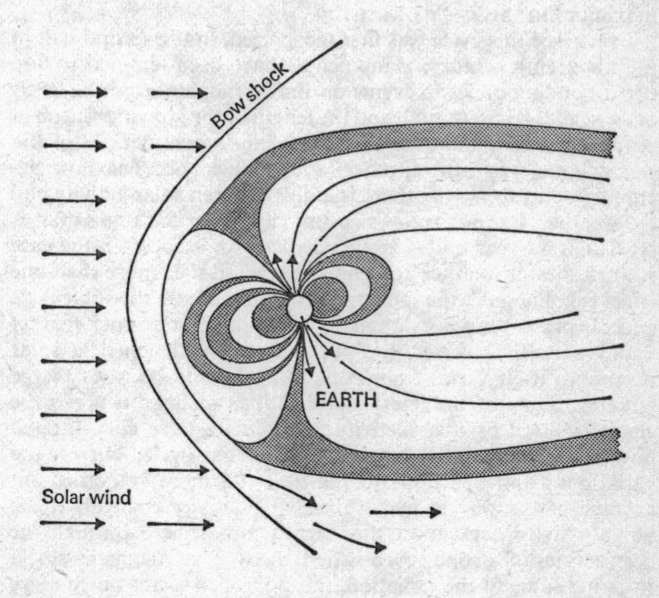

Figure 4.2 The Earth in space, showing the magnetosphere (not actually a sphere but a stretched out ellipsoid, moulded by the solar wind). Close to the Earth, our own magnetic field dominates the region above the atmosphere proper where charged particles (ions) are trapped in the Van Allen Belts (shaded). Further out into space, there is a constant wind of charged particles 'blowing' outwards from the Sun. Between these two regions, the boundary between the Earth's magnetic influence and the solar wind is marked by a bow shock – the equivalent of the bow wave of a boat moving through the water. Along the bow shock, particles from the solar wind penetrate the Earth's sphere of influence and may eventually funnel down into the atmosphere near the polar regions; trailing behind the Earth as it moves through the solar wind the long tail of the magnetosphere constantly leaks particles away out of the Earth's sphere of influence and into the solar wind which blows on out into interstellar space.

wind is stronger and more gusty – at the time around solar maximum in this cycle of activity.

Ever since this cycle was first recognized, in the second half of the nineteenth century, many people have been tempted to find corresponding cycles in events on Earth, ranging from the levels of African lakes to rainfall and the length of the growing season of crops in Britain, and even to cycles of the stock markets of the world. One way and another, enough evidence has now accumulated to show that there is a link between solar activity and the weather, but not an overwhelmingly important one as far as particular 11-year cycles are concerned. As is so often the case with studies of weather and climate, we find that more than one effect is going on at the same time, and in this case the solar cycle effect is present but not dominant (especially not in stock market fluctuations!). However, the peak of activity reached at solar maximum itself varies from cycle to cycle, and by a very large amount. Some of the latest investigations of links between the changing level of solar activity and climate have concentrated not on the variations over each individual cycle, but on the variations from one cycle to the next – a more smoothed-out average of the way in which the Sun's activity changes. If the Sun's activity does affect the Earth's atmosphere through the destruction of ozone, even when there is a magnetic shield stopping some of the radiation, this approach seems much more likely to prove fruitful than looking at year-to-year variations because, as Dr Reid's group have shown with their computer models, although the effect on ozone is most pronounced a year or two after an outburst of solar activity it takes about ten years for the ozone level to return to normal. So any search for this kind of solar influence affecting climate should really look at averages over ten years or so, smoothing out the detailed fluctuations within individual solar cycles of activity.

So – are there any signs of ozone fluctuations in the stratosphere, produced by the changing level of solar activity and changing strength of the solar wind? Evidence reported by Dr Ronald Angione and colleagues in 1976 suggests that there is indeed such an effect, and that natural fluctuations in ozone content of the stratosphere in this century have certainly been bigger and longer-lasting than any influence of man's activities. The evidence comes from a long-term monitoring programme in which data were obtained by the Smithsonian Astrophysical Observatory from sites in southern California and northern Chile during the first half of this century. The observations turn out to be uniquely valuable because they were made in the 'Chappuis Band', a part of the electromagnetic spectrum between 0·5 and 0·7 micrometres wavelength. The significance is that the absorption in this band happens to be unaffected by temperature, which does affect monitoring of ultra-violet radiation from above the stratosphere at other wavebands. Because of this, the influence of ozone in absorbing radiation passing through the stratosphere can be determined in a fairly straightforward way, with effects of absorption by other substances subtracted out from the calculations using data obtained by monitoring at different wavelengths.

The result is clear-cut and impressive. To quote Dr Angione's team, 'at both sites, total ozone amounts commonly showed variation of as much as 20 to 30 per cent on time scales ranging from months to decades'. And this immediately suggests a link with climatic changes.

Because of its position in the electromagnetic spectrum at a region where the Sun is emitting a lot of energy, the Chappuis Band absorption by ozone alone absorbs something like 2% of the total solar energy reaching the stratosphere. A fluctuation of 25% in the ozone concentration can alone produce a fluctuation of 25% of 2%, or ½% in the amount of energy getting through to the ground. And that, as we have already seen, could alone be enough to produce a change in the mean temperature at ground level of up to half a degree Centigrade. Furthermore, because the slanting rays of the Sun travel a greater distance through the ozone layer on their way to high latitudes than on the way to tropical regions, the effects will be most pronounced in those high latitude regions where the atmospheric circulation is most susceptible to outside influences.

We have not, as yet, proved that these fluctuations in ozone concentration in the stratosphere are produced by the changing level of solar radiation penetrating through the shield of the

Earth's magnetosphere. But the probability of such an effect being real is certainly increased by a study of the way in which the apparent brightnesses of some of the other planets in the solar system vary over the solar cycle, reported by Dr G. Lockwood, of Lowell Observatory in Flagstaff, Arizona.

The observations, of the planets Uranus and Neptune and the moon Titan, show a variation in brightness following a cycle in line with the changing level of solar activity in recent years. These astronomical bodies seem to be brighter at solar minimum, by an amount which suggests on the most naive interpretation that the Sun's brightness varies by about 2% over the solar cycle of activity. The point is, of course, that planets and moons do not shine in their own right, but because they act as mirrors, reflecting the Sun's light. Any change in the brightness of the reflection ought to mean that the brightness of the Sun has changed – unless the nature of the mirrors varies in some way.

At present, this second alternative seems to be the one favoured by astronomers. After all, the atmosphere of a planet is variable, and not to be compared with the permanent silver coating on the back of a glass mirror. If the nature of the solar wind of particles blowing out to the planets varies – as we now know it does – then it could produce changes in their atmospheres which make them more or less reflective. If solar wind particles can affect ozone in the stratosphere of the Earth, there is every reason to expect them to influence the photochemical reactions which go on high in the atmospheres of Uranus, Neptune and Titan. So, perhaps these bodies are simply more shiny at solar minimum than at solar maximum, when the stronger solar wind has the effect of dulling the mirrors by a small amount. If so, here is circumstantial evidence, at least, that the solar wind affects planetary atmospheres by an amount corresponding to an influence of up to a couple of per cent of the amount of energy reaching them.

Another explanation which could apply equally well to the problem of links between solar activity and climate is that while the Sun's total output of energy over the entire solar cycle of activity remains constant, the amount radiated at different electromagnetic wavelengths changes slightly between the time of solar minimum and solar maximum. In other words, the 'colour' of the Sun's radiation may change, and in particular there might be more or less radiation arriving at the stratosphere in the Chappuis Band, and correspondingly less or more radiation arriving in some other band. Again, this could explain the apparent changes in brightness of Uranus, Neptune and Titan, since their atmospheres may well reflect more effectively in some wavebands than in others. And again, of course, such a change would affect the amount of radiation which gets through the ozone filter to the ground here on Earth.

The third possibility for explaining Dr Lockwood's results is the one that most astronomers shy away from instinctively. Perhaps the total energy output of the Sun really *does* change by as much as two per cent over the length of an individual solar cycle. For a trained astronomer to suggest such a thing is almost heretical – our Sun has long been regarded as the archetypal 'typical', steady star, glowing with unwavering intensity. But in recent years several doubts concerning our understanding of the workings of the Sun have arisen, at the same time that research into climate and weather has hinted at the significance of Sun–Earth interactions, perhaps involving variations in the Sun's output. Now, only a trained astronomer can appreciate just how much uncertainty there is surrounding the workings of the Sun, and admit that perhaps the Sun is not a 'normal' star. The most dramatic fact, seldom appreciated even by meteorologists, is that in fact we have no observations of the Sun which show that it does really burn steadily to within 2% of its average brightness. Observations made through the Earth's atmosphere are useless for such measurements, since as we now see the filtering effect of the atmosphere itself varies measurably. Observations above the atmosphere, made from space probes, will ultimately settle the question one way or another – but although we do just about have the technical expertise these days to build a satellite which could monitor the total energy output of the Sun to an accuracy better than 2%, no such satellite has yet been planned, built or launched – and when it is, we will then need observations over a whole 11-year cycle before we can use the results to answer this question about solar variability!

Meanwhile, we have to take the prospect seriously – indeed, my own belief is that when such a satellite is available it will show small variations in the output from the Sun. And if we now look at some details of the known Sun–weather interactions, together with the puzzles astrophysicists now have to tackle in order to explain recent observations of the Sun, we will see just how seductively persuasive the idea that solar variations provide the best guide to climatic changes that will be important in our lifetimes really is.

Chapter Five

Flickers in the solar furnace

The mass of circumstantial evidence which relates changing climatic factors on Earth to the changing level of activity over the solar cycle is too large to examine in detail here. Most of the evidence is a bit like candy floss – it looks impressive, but reduces to a tiny fragment as soon as you try to get your teeth into it. The reason for this, it seems to me, is that although there are real solar cycle effects on the changing climate, the amount of solar influence that fits exactly with the roughly 11-year cycle is much smaller than the total of all the other variations in the climatic system, and certainly has little, if any, more significance than the changes produced by the 'random noise' effects mentioned earlier. Because of this, the evidence from one particular study may look impressive over the history of one or two solar cycles, but generally the sceptic can pick out another run of solar cycles in which this particular effect cannot be seen at all. In order to convince ourselves that the solar cycle influences on climate are real (if small) we have to resort to quite sophisticated statistical tests, of the kind developed by electrical engineers in order to pick out a weak 'signal' from a great deal of random noise. This approach, although requiring some mathematics, is quite clear and unambiguous; it may not have a great deal of climatic relevance on a broad scale, since the effects are indeed small and presumably average out over one cycle of solar activity. We are not especially concerned about small variations which go up and down over 11 or 22 years, but with longer-term trends that make bigger changes over 50 years or so. But the new statistical evidence for a solar cycle 'signal' in climatic variations on Earth is of the greatest importance because it tells us something about the physical links between the Sun and the Earth.

If some kind of solar variation produces even small changes on Earth within the timespan of each 11-year cycle, then it will hardly be surprising to find that bigger solar variations over longer timespans produce much more important climatic changes.

Dr Robert Currie, for example, reported in 1974 that appropriate statistical tests carried out on temperature records from various sites in the North American continent, each of them covering at least 60 years and with some records going back well into the nineteenth century, show a 'solar cycle signal' with temperature variations of about 0·1°C occurring with an average period of just under 11 years. Equally important, Currie found that there is no evidence of any other period between 9 and 25 years long (the range covered by his tests) – so the *only* regular effect in all these records is the one which can be related to the solar cycle. On its own, this kind of information does not tell us whether the heat output of the Sun – the quantity of radiation – varies over the solar cycle, or whether the nature of the radiation and solar wind – the quality of the radiation – changes and affects the Earth's atmosphere to produce the solar cycle signal. But when the mathematical tests show that the signal is there, we can look at some of the 'folk lore' about sunspots and the weather with a slightly less critical eye, taking them – if not quite at face value – as a genuine indication of some basic solar influence on climate and weather.

A good example of the value of this hindsight approach lies in the farming records of the Farrar family in Devon. John Farrar, one of the partners in the family farming business, came across the idea of a solar cycle influence on weather and harvests in 1974, a year when many farmers and growers in the UK were having a particularly bad time with the weather. Looking back into his own records and those of the family before him, he found previous bad years in 1963 and 1952, which immediately hinted at an 11-year pattern and encouraged him to work out some statistics which, if less sophisticated than those of researchers such as Dr Currie, are quite convincing in their own right. Diaries kept on the Devon farm since 1947 show very clearly how the length of the summer 'growing season' has changed in the past 30-odd years. The herd of some 100 cows kept on the farm is put out to graze each year as soon as there is sufficient grass, and brought in again in the autumn as soon as it becomes necessary to start feeding them silage or hay. These dates, which mark the extreme limits of the growing season of the grass as unambiguously as any other records, can be used to construct a graph (Fig. 5.1) comparing the length of the grazing season with changes in the number of sunspots. Although there is some 'noise' in the pattern, the overall changes in grazing season do seem to follow a pattern in which bad years for farmers in Devon come shortly after the minimum of solar activity, while good years come just after periods of high solar activity. For those farmers with the courage to take advantage of it, this offers a practical advantage in the next few years.

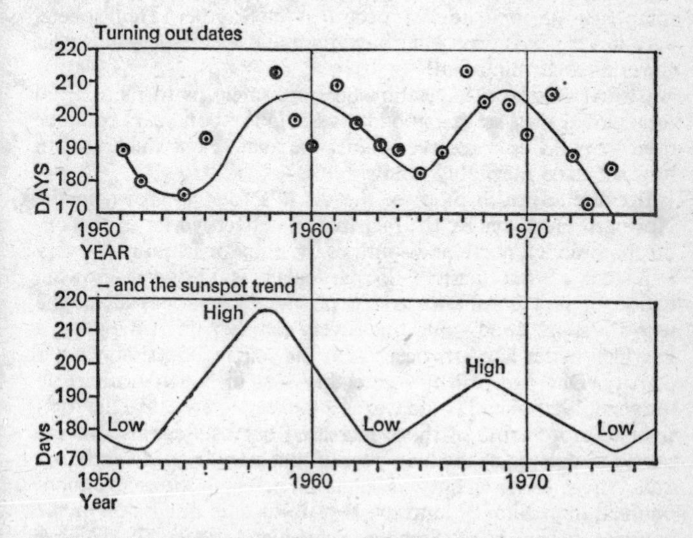

Figure 5.1 Number of days spent out at pasture by the Devonshire dairy herd of the Farrar family in each of the past 25 years, compared with the number of sunspots observed each year. Based on figure in John Farrar's article in *Farmer's Weekly*, 28 February 1975.

In the early 1970s, farmers in England were increasing their stocks of cattle after the successes of previous years, just at a time, we now see, when the solar cycle effect was working against them – with disastrous results. At present, there is an air of caution after those disasters – but in fact now is the time to build up stocks for the expected good years at and just after the next maximum of solar activity, which is due at the beginning of the 1980s. And this kind of effect is not just restricted to Devon, or to the United Kingdom. Several researchers have now found similar links between the solar cycle of activity and agricultural production around the world; a team headed by Dr J. W. King, at the UK's Appleton Laboratory, has found in particular that while harvests over the whole Northern Hemisphere of our planet follow this pattern of increased yield near years of solar maximum, just the opposite effect occurs in the Southern Hemisphere, with lower yields near solar maximum and increased yields near times of solar minimum!

Clearly this kind of relationship could help us to plan ahead on a global scale so that good harvests in the best years could be used to build up a reserve against the lean years which we can now see must inevitably follow.

Because so much more of the world's food is grown in the Northern Hemisphere, the fact that the effect works in the opposite direction north and south of the equator is not really very important. What matters, in particular, is the grain-growing region of North America which provides just about all of the world's 'spare' food – and this is very clearly part of the pattern in which better harvests occur near the years of maximum solar activity. Dare we plan for better harvests in North America in the early 1980s – and could we, as a global society, have the foresight to store some of those increased harvests against the expected bad years of the later part of that decade, when the solar cycle effect turns against us again? These questions are more political than climatic, and the best up-to-date discussion of the political puzzle is in Stephen Schneider's book *The Genesis Strategy*. But before we look more closely at what causes these changes, one other solar effect with clear agricultural implications must be mentioned.

Drought periods in the central United States seem to be correlated with the solar cycle of activity, and especially with the 'double' sunspot cycle which averages 22 years in length. Just as the '11-year' cycle itself can actually range from about 9 years to 12 or 13 years in length, the 'double' cycle can be anything from 18 or 20 years to 24 or 26 years long; within this range of accuracy, the coincidence between the US droughts and the double cycle is impressive, with only one notable drought year (1845) outside the pattern. Drought peaks at St Louis, Missouri, occurred in 1838, 1845, 1854, 1872, 1895, 1914, 1931 (and the notorious 'dust bowl' years of the period) and 1955, a run which, regardless of the cause of the droughts, pointed strongly at the possibility of drought hitting the corn and wheat belts of the US some time in the middle of the 1970s – a drought which certainly arrived in some parts of those regions. Again, on this picture the next period of prolonged good weather (for agriculture in the US) should be in the 1980s. Perhaps the most important lesson here is the warning that this should not encourage complacency. As harvests improve over the next few years it will be all too easy to be lulled into a false sense of security, so that the bad harvests of the 1990s, coming at a time of greatly increased population, produce even worse shocks to the world system than the bad harvests, droughts and famines of the 1970s. Things may get better before they get worse, but that worse is still to come.

But is the 22-year cycle a sensible thing to blame on the Sun? It turns out that it is, because as well as the simple increase and decrease in numbers of sunspots over the 11-year cycle the Sun also undergoes an overall change in magnetism, with effects on the magnetic polarity of sunspots, that takes about 22 years to complete. To astronomers, it is this 22-year cycle that is the fundamental cycle of solar activity, and the 11-year sunspot cycle is just one effect produced by the more basic changes over the 22-year cycle. What we see from this is that it is not the sunspots themselves which affect climate on Earth; the variations in sunspots and the variations in climate are both effects produced by some more fundamental change in the Sun. All the flaring outbursts on the Sun, its spots and the changing solar wind are symptoms of some deeper change. The deeper change itself may well affect the climate of the Earth in the longer term – 22 years and more – but even though sunspots may not affect us directly, one of the symptoms, the changing solar wind, certainly produces its own effects in the terrestrial weather.

The wind that shakes the world

Since the early 1950s there has been a growing weight of evidence that effects linked with changes in solar activity could produce specific, measurable changes in weather systems on Earth at high latitudes. In particular, it seemed from many studies carried out by Professor Walter Orr Roberts and colleagues that the depressions (lows) sweeping across Alaska after the occurrence of bright auroral displays – themselves following hot on the heels of solar flare outbursts – were deeper and more intense than the usual run of such depressions. For most of the past quarter century, this kind of evidence remained in the realm referred to derogatorily by some scientists as 'stamp collecting'. The pattern could be seen, time and again, but there was no explanation, and without an explanation the pattern remained largely ignored.

The reason for the lack of explanation was quite simple. Until very recently we knew hardly anything about what goes on in space between the Earth and the Sun, with no observations of the existence of a solar wind of particles blowing out past Earth, let alone any information about how the wind changes as the activity of the Sun changes, and as the Sun rotates. The more that has been discovered about the solar wind, using space probes, the more easy it becomes to understand and explain the results Roberts has been pointing out for more than two decades; and with more and more 'stamps' in his collection, by the early 1970s it also proved possible to begin to carry out the statistical tests which provide the ultimate mathematical proof of the existence of a real Sun–weather correlation of this kind. The better physical understanding of solar–terrestrial links, and the improved statistics, have now combined to provide a convincing explanation of why individual solar flares can affect the weather. Furthermore, there is at least a hint here that some of the 'solar cycle' influences on weather may simply be the averaged out effect of many solar flares occurring near the years of solar maximum, and a few occurring in the years of the quiet Sun.

In 1972, Professor Roberts and Dr Roger Olson, of the University of Colorado, reported a study of the effects of solar activity on weather systems at a height in the atmosphere where the pressure drops to 300 millibars (roughly 9 km altitude) in the region of northern North America and the North Pacific. The effects on the weather systems were linked with changes in the magnetic field of the Earth – geomagnetic activity – which can be triggered off by the arrival of bursts of charged particles from the Sun. The troughs (depressions) moving into the Gulf of Alaska a couple of days after a steep rise in this geomagnetic activity are bigger than average, as the stamp collecting had already shown, but now the chain of cause and effect had been lengthened and strengthened by the knowledge that there is a sound physical explanation for the geomagnetic changes in terms of changes in the solar wind.

The next step, obviously, was to find out more about those solar wind changes, and for this the Colorado team worked with Professor John Wilcox and his colleagues at Stanford University to pin down the solar–terrestrial weather link.

Spacecraft observations show not only that there is a wind of charged particles blowing outwards from the Sun, but also that there is a structure within this wind, a structure reflecting the structure of the solar magnetic field. From the viewpoint of all of us here on Earth, the structure is like fat slices of pie dividing up the Earth's orbit, with the magnetic field polarity either towards or away from the Sun in each pie slice, and with opposite polarity in slices that lie next to each other (Fig. 5.2). The boundaries between the sectors are very narrow, and usually there are four slices, each roughly covering an angle of 90° centred on the Sun. The Earth moves around the edge of this pie once in a year, but this is a very slow movement compared with the rotation of the Sun itself, which spins the whole pie structure around and past the Earth, so that we see one complete rotation of the whole pattern every 27 days. In that period, typically four sectors, and four sector boundaries, pass across the Earth in its

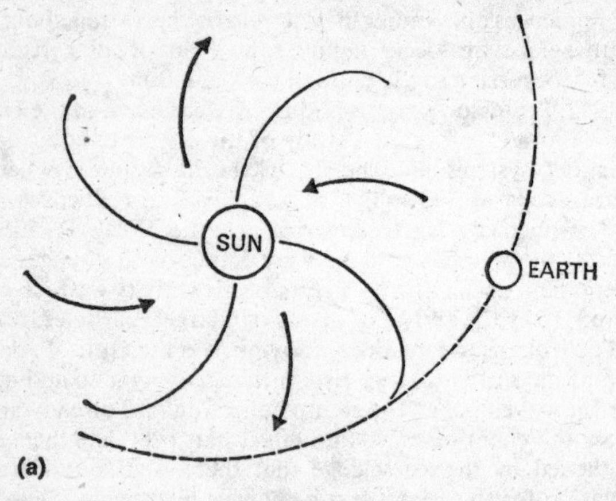

Figure 5.2a The 'pie' structure of the solar magnetic field. Arrows show average direction of field, which reverses at the marked boundaries. The dotted line is a schematic indication of the Earth's orbit (cf. *The Jupiter Effect*, p. 94).

orbit. When a sector boundary crosses the Earth, and the Earth's magnetosphere, it produces a sharp jolt in geomagnetic activity one or two days later; the Colorado–Stanford team used this marker, the passage of a sector boundary, in their weather studies.

Using an index of the strength of weather systems called the Vorticity Area Index, or VAI, they were soon able to put the stamp collecting on a much firmer basis, with a description of the typical effects of the passage of a sector boundary. First, the VAI increases by about 10% within the next two or three days. This result is quite unambiguous, at least for the winter months, and was alone enough to establish beyond doubt that the solar wind affects our weather. Indeed, Professor C. O. Hines and Dr I. Halevy, of the University of Toronto, who were so sceptical of the results that they carried out their own statistical tests, reported ruefully in the journal *Nature*, that they had succeeded only in convincing themselves that the effect was real! When the

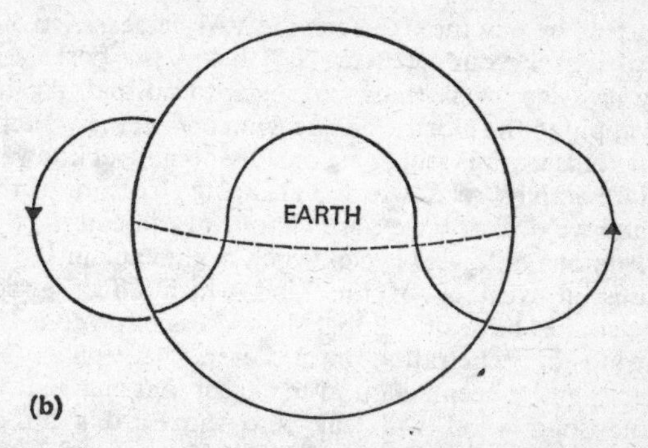

Figure 5.2b The solar 'magnetic equator' is like the seam on a tennis ball. The magnetic field comes across this 'seam' as indicated. The field points 'out' or 'in' along the rotational equator (dotted line) – which creates the 'pie slices' shown in the previous diagram.

sceptics begin to be converted, progress usually follows rapidly, and this case was no exception. Before long, Olson and Roberts, with Dr C. S. Zerefos, were reporting results of a new study tying in flares on the Sun to the changes in geomagnetic activity and VAI on Earth.

Earlier results had shown atmospheric changes at both 300 and 500 millibar levels in the atmosphere and a decrease in VAI one day after the crossing of a sector boundary, with a recovery of the VAI back to normal before four days had passed. But the crossing of a sector boundary is only a reference point within the pattern of changing solar and solar wind activity, and the researchers at Boulder sought the explanation of why the solar wind should change by looking at the occurrence of flares on the Sun – great outbursts of activity which are more common when there are more sunspots, and which are often associated with groups of sunspots. Once again, a very clear pattern of events emerges from the study, a pattern which takes about six days to complete. By day one or two after a flare on the Sun, the VAI has increased to 5 or 10% above the 'normal' background level; by day two or three the geomagnetic storm associated with the flare has started; by day three or four the VAI decreases to 5 or 10% below the background (10 or 20% below the post-flare peak) and by day five or six things are back to normal. Again, these results apply to the North Pacific region and to the winter half of the year. This work is still going on – we do not yet know how the solar flare activity relates to the changing structure of the solar wind, and we do not have observations over a complete 11-year cycle (let alone a 22-year cycle) which might explain some of the variations in weather of the kind which affect agriculture. Satellite studies have shown that the success of protons from the solar wind in penetrating into the magnetosphere is quite significant, and it seems natural to expect that charged particles from the solar wind will leak into the Earth's atmosphere through the 'holes' in the magnetic shield near to the magnetic poles. It should be no surprise, then, to find this kind of solar effect on the weather most pronounced at high latitudes, near the poles. And those are just the latitudes where small changes in the overall circulation of the atmosphere can have a profound effect on climate in the critical grain-growing regions. We also know now, again from spacecraft studies, that the solar wind is more 'gusty' in the years around solar maximum activity, as we would expect with the flares producing extra bursts of particles, or solar squalls, to sweep across space and shake the Earth. Ideas about how any of this affects the weather are still speculative, but what Professor Roberts and his team have now done is to make such speculation respectable.

We don't yet know how the solar–weather effect works, and we cannot predict it over the years ahead. But we do know there is a real effect, and, as with the statistical tests of solar cycle signals in temperature, rainfall and so on, the fact that we know there is a real effect encourages us to accept some of the ideas linking solar activity and climate which have long been regarded as slightly disreputable. We have one possible explanation of all the solar cycle effects, that they are merely the averaged-out effect of many solar flares. This qualitative explanation has the great merit that it also explains why the solar cycle effects are not exactly regular – changes in the weather don't exactly follow sunspot numbers because they are linked, not with sunspots, but with the fluctuating flare activity of the Sun, and with the squalliness of the solar wind. But when we look beyond the 11- and 22-year cycles this effect is seen to be insufficient. There must be something else at work to explain the difference between, say, the 1950s and the 1680s. And that could be where genuine quantitative changes in solar output contribute to the pattern of our changing climate.

Longer-term effects

Although we only have records of direct observations of sunspots going back for a couple of centuries, there are many second-hand indicators that have been interpreted as giving us some guide to the changing level of activity of the Sun over the past few thousand years. The best of these in some ways are historical mentions of auroras, because we now know that bright, extensive auroral displays are produced when the Sun is most active, sending many bursts of particles across space to Earth in the gusts of the solar wind. There are some direct mentions of sunspots in both Greek and Oriental records, and one way and another some idea of at least which years were the years of solar maximum activity can be worked out over a period going back well before the birth of Christ. But there is another effect of the changing level of solar activity which is at least as important as the actual length of each solar cycle, or the date of each solar maximum year. This is that the strength of each peak of solar activity varies considerably from one cycle to another, as is very clearly shown in Fig. 5.3, which is based on the data prepared by Dr M. Waldmeier quoted by Dr John Eddy in a paper in *Science* in 1976.

Looking just at the height of each peak in the solar activity graph, we can see that some regular effects are present which cover periods much longer than 11 or 22 years. Once again, we have to turn to the statistical tests used to pick out signals from noise in order to find the basis of these ups and downs in the strength of succeeding solar cycles, and unfortunately this time the evidence is too scanty for the statistical tests to be completely

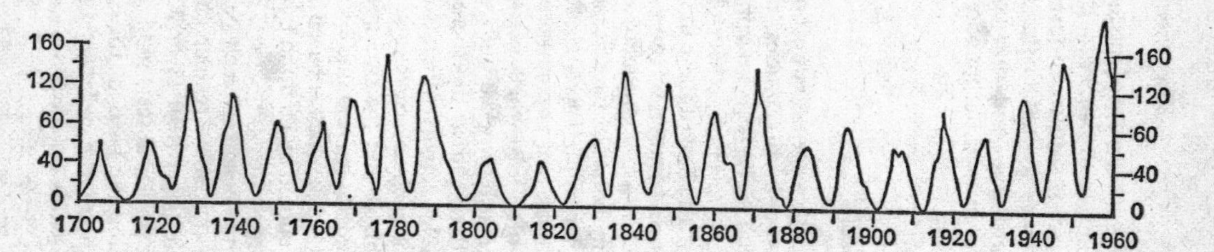

Figure 5.3 Changing numbers of sunspots observed in the years from 1700 to 1960. Notice that as well as the roughly 11-year periodic variation the *size* of each sunspot peak varies from cycle to cycle. Based on data of M. Waldmeier, as reported by John Eddy.

successful. But even with this limited information, a 'supercycle' with a period of between 80 and 90 years seems to be present – and with some of the second-hand information added to extend the pattern into the more distant past there seems to be some evidence of a pattern of activity recurring after about 180 or, in round terms, 200 years. Just as the 11-year and 22-year cycles are themselves seen as different aspects of one basic variation, it is fairly clear that if the two longer-term effects are real they could also be different aspects of one overall variation, with the roughly 90-year 'cycle' half the roughly 180-year 'cycle'. Of course, any attempt to interpret solar variations in this way needs more than even the most reliable eye-witness accounts of how bright the aurora were in, say, AD 1154. But fortunately there is a record which can now be interpreted to give a guide to the changing strength of the solar wind – the record of radioactive carbon-14 found in all living things.

Carbon-14 is produced in the atmosphere when the energetic particles of the solar wind interact with nitrogen atoms. More carbon-14 is produced when the solar wind blows stronger (or when the shielding magnetosphere is less effective, but that is not an important consideration when we are only looking at the past few hundred years), and like ordinary carbon-12 most of it ends up in carbon dioxide, which is absorbed by plants, in particular, and remains in wood and other substances as a direct record of how the level of solar activity has changed. Wood is especially valuable, because the age of a particular piece of wood can be found by counting the tree rings in it – and a tiny sliver of wood from within one ring is quite enough for the tests which show the proportion of carbon-14 to carbon-12 present in the carbon dioxide of the air when that wood was being laid down.

This kind of study shows that periods of prolonged high solar activity are associated with warm winters, while periods of quiet Sun, in which even the peaks of solar activity in each 11-year cycle are low, produce cold winters in Europe and elsewhere across the Northern Hemisphere. The evidence also shows the 90- and 180-year cycles, as well as even longer-term trends, which might be important to long-term climatic changes – even, perhaps, to Ice Ages – but need not concern us greatly in our search for an understanding of the climate we can expect in the next half century.

In order to predict how the long-term changes in the solar cycles are likely to affect us, we can just extend the known cycles forward – but that is a little unsatisfactory. It's much better if we can find even a rough guide to what is causing those changes, and use that to make a firmer base for our predictions. There seems to be only one candidate to explain how the Sun can get stirred up in a long-term cycle of activity lasting roughly 180 years, and that depends on the movement of the Sun and all the planets of our Solar System around their common centre of mass.

This represents a new step for most people. The idea that the Earth is the centre of the Universe has long been superseded and we know that even our Sun is just one insignificant little star among thousands of millions of stars in our Milky Way Galaxy. But surely the Sun is at least the centre of the Solar System? Alas, no. The Sun is very much more massive than all the planets put together, but the spread of the planets across space helps to shift the 'centre' of the Solar System considerably relative to the centre of the Sun. If we try to balance two objects with very different mass on an ordinary beam, we find that the point of balance lies very close to the more massive object. With just the Sun and one planet – perhaps Jupiter, the biggest in our Solar System – a similar effect would occur. The centre of mass of the system would be a point close to the Sun, but not quite at its centre. And both the Sun and Jupiter would orbit around this point – *not* about the centre of the Sun. The same thing occurs with the Moon and the Earth. We think of the Moon as orbiting around the Earth, but in fact both the Earth and Moon orbit about the common centre of mass, a point which lies so far below the surface of the Earth that for most purposes we can pretend quite happily that the Moon really does orbit around the centre of the Earth.

But the Sun has a family of nine planets, all at different distances from it, and all with different masses, plus various bits and pieces of cosmic junk held by the Sun's gravity in a variety of orbits. This complicates the picture, not least since all the different objects orbit around the common centre of mass with different periods – their 'years' are different lengths. But by looking only at the main influences, the Sun and planets, astronomers can work out roughly what happens as all these different motions shift the centre of mass of the whole system about. If we return to our conventional picture, and pretend that the Sun is fixed in space while the planets move about it, we can calculate how the centre of mass of the system seems to move. It follows a looping path which carries it across the interior of the Sun and sometimes outside the Sun altogether, in a complex pattern of clover leaves and figures of eight as the planets dance around. But this is, of course, a false picture. In reality, the centre of mass is fixed (if anything is) and it is the *centre of the Sun* that follows the complex, looping path around this true centre of the Solar System. It's as if the whole Sun was on the end of a short piece of elastic being swung round and round, while the elastic stretches and contracts rhythmically. Since the Sun is a fluid body, with the surface layers in particular being made up of something comparable to what we think of when the term 'gas' is mentioned, this extraordinary motion is quite likely to have an effect on the surface features that we can see. After all, if you tied a bucket of water to an elastic rope and swung it about, with the odd jiggle thrown in for good measure, you wouldn't really expect the surface of the water in the bucket to stay smooth and unruffled.

So, on this picture it would be a surprise if the Sun did not change, at least on the surface, as the planets tug it first one way and then the other. What sort of an effect might be produced? Theory alone cannot take us any further – but we can say how whatever the effect is should vary. The biggest planet in the Solar System, by far, is Jupiter, so Jupiter should produce the biggest individual contribution to the solar jiggling. Jupiter takes just about 11 years to complete one orbit around the Sun, producing a jiggle with just this period, strikingly close to the basic 11-year period of solar activity. This coincidence has been noticed before, but on its own raises the puzzle of why solar activity does not vary *exactly* in line with Jupiter's motion. Now, we can guess that this is because we have not yet allowed for the effects of the other planets. When these are all added together, it turns out that the whole cycle of movements produces an effect with varying periods around 11 years, and that the pattern as a whole repeats with a period of just about 179 years – a result so striking that I find it impossible to believe that the 180-year cycle in solar variations can be attributed to any other cause. But as an astronomer I must look for some convincing evidence that the Sun's appearance really is likely to be susceptible to this kind of external influence, before accepting the coincidence at face value and using it as a basis for climatic predictions.

Can the Sun vary?

A few years ago, the idea that our Sun could be a variable star, even on the relatively modest scale needed to explain climatic variations of recent centuries, would have seemed outrageous to most astronomers. The Sun was regarded as the classic example of a stable, unvarying star – although admittedly the solar cycle of activity might have caused a few qualms about this broad generalization. But in the late 1960s and through the early 1970s just about every new test applied in the study of the Sun came up with a 'wrong' answer according to the standard picture. The first cracks in the picture came with the search for solar neutrinos, particles that should be produced in copious quantities by nuclear reactions going on in the centre of the Sun, if the standard picture is right. The neutrinos have not yet been found, after years of searching, and the only sensible explanation is that they are not being produced in the Sun, that it has gone 'off the boil' temporarily, and cooled down in the middle by perhaps 10% compared with the astrophysicists' standard models. More recently, news has come that the latest observations of how the surface of the Sun moves show that the whole great mass of gas wobbles like a blancmange, pulsing in and out with a period of 2 hours 40 minutes, with the wobble reaching a maximum velocity of about 2 metres a second. Just as earthquakes can give us information about what is happening inside the Earth to make the surface shake, so this kind of 'starquake' can tell us something about the processes going on inside the Sun. In the words of the team that discovered this wobble, the 2 hour 40 minute period is 'quite consistent with the observed absence of any appreciable neutrino flux from the Sun', and suggests independently that the middle of the Sun is about 10% cooler than our standard models require. The period cannot be explained so simply if the Sun really is like our standard models.

As if opening a floodgate, the announcement of this discovery soon brought news of other variations, perhaps of a similar kind, in other stars. Dr Norman Walker, of the Royal Greenwich Observatory, gave details of these in 1976 in an interview with Ian Ridpath, reported in *New Scientist*. A four-year study of some bright stars (each about 20 times as massive as our Sun) has revealed evidence of oscillations with periods in the range from 40 minutes to an hour or so, but with amplitudes 10 to 100 times greater than the newly discovered solar vibrations. As Walker puts it, 'whatever property the Sun has, we find some stars have much more of it' – which is not, after all, so surprising for stars so much more massive than our Sun. Our Sun may not be what we used to think of as a 'normal' star, but nor is it unique – which would have been a very worrying discovery!

If we are to explain these peculiarities at all, the answer probably lies somewhere in the layer of the Sun that is dominated by convection – the region, well out from the centre, where heat from below causes the fluid to rise outwards before it cools and falls back again towards the hot interior. A lot of the deeper layer which does not take part in convection (according to those standard models, which now look slightly less than ideal) must be very close to the edge of convective instability, and if given a push might begin to take part in this convective flow, fairly quickly transporting a lot of heat outwards from the centre of the Sun.

If this happened, for whatever reason, the centre of the Sun would abruptly cool down, a lot of the convection would stop, and then the nuclear furnace at the heart of the Sun would gradually build up again towards full power. And I mean gradually – the time it takes the Sun to regain equilibrium, once it has been disturbed in such a way, is likely to be at least ten *million* years, even though the disturbance which took it off the boil might only take a few thousand years. It could very well be that something happened to upset the Sun's convection some time in the past ten million years or so, and that it is still struggling to get back to 'normal'. Perhaps even the whole of the recent Ice Age might be directly attributed to changes occurring as a result of that struggle – but such speculations are outside the scope of the present book.

There is at least one plausible explanation for why the Sun should be disturbed just now, an idea which gained some favour a couple of decades ago but was then dismissed, only to be revived in modern form by Professor W. H. McCrea, of the University of Sussex. As the Solar System moves around our galaxy, every few hundred million years it runs into relatively thick clouds of interstellar dust and gas; the effect of this material falling on to the Sun might be enough to trigger some kind of instability, which I envisage as follows.

If we liken the normal, stable convecting Sun to a pot boiling on a stove, the arrival of the dust acts as a kind of 'lid', allowing the temperature within to build up and, if anything, suppressing the convection. When the Sun and Solar System emerge from the dust, however, it is as if the lid is removed. Suddenly (by astronomical standards) convection runs riot for a brief spell and produces a dramatic cooling inside the Sun – an effect which we can envisage as like taking the lid off of a pressure cooker. Such an effect would certainly do the required trick – and we last emerged from a dust cloud of this kind only about 10,000 years ago. If that is when the interior of the Sun cooled, we may be 10 million years away from normality even now.

That is just one idea, which I particularly like. Other astrophysicists have come up with other ideas about how the Sun could go 'off the boil' temporarily, now that they have been prompted by the discoverer of the Sun's various peculiarities, and it sometimes seems remarkable in the light of this newly-fashionable school of thought that the Sun should be in a stable state at all. But certainly it is clear that even when it is stable the Sun is not far from a situation of convective instability, and that any disturbance – perhaps even a build-up of 'nuclear ash' in the centre – may push it over that edge. All we need is some such trigger some time in the past few million years – a few million years in which we know the Sun has been ploughing through a particularly dusty bit of the Milky Way Galaxy, which is why I favour the idea of a dust 'trigger' to explain the Sun's present unnatural state.

All this may seem far removed from the puzzle of how climate is going to change between now and the end of the twentieth century. But if our Sun really is at present in a delicate state, with a sensitive convective region still trying to return to normal, and all the time it is being swung about the Solar System centre of mass by the changing pull of the planets, it is hardly surprising that we see the ripples that result as a changing level of solar activity, including sunspot variation. With this new picture of the Sun as a rather seedy, off-colour star subject to constant pushing and prodding it seems quite reasonable to use the 180-year, 90-year and 22-year cycles, as well as the 11-year cycle, as a secure basis for climatic forecasting at present. Whether the effect on climate is produced by real changes in the heat output of the Sun, or by changes in the nature of the solar wind, or by qualitative changes in solar output which alter the transparency of the ozone layer does not matter. The effect is real and the results are the same – results which we can now look at more closely if we come back from the astronomical perspective of millions of years and galactic distances to the past few hundred years' history of one small planet.

Chapter Six

Forecasting the flickers – and the climate

The solar flickers that are important for climatic prediction on the scale of a human lifetime are the flickers which make the peaks of individual solar cycles themselves go up and down. Very long-term solar instabilities pose potential problems, but too remote to be of immediate concern. Variations within each solar cycle pose immediate problems, but only small ones of the kind which farmers have become used to – part of the accepted year-to-year variability of climate. The intermediate variations are both the most puzzling and the most significant for human planning today.

Evidence of variations in the Sun's output (the solar 'constant', now better referred to as the 'solar parameter') has been around for many years, without having been taken very seriously. Professor Lamb has summarized the arguments about the constancy of the solar parameter in his definitive book *Climate: Present, Past and Future*, and the evidence includes, for example, observations made by Dr C. G. Abbot at the Smithsonian Institution between 1920 and 1955. These show, on the face of it, a systematic variation within the solar cycle of about ½%. Following Dr Abbot's retirement in 1955, these observations were, unfortunately, discontinued. But his basic findings have been confirmed by measurements made with instruments flown on balloons in the stratosphere between 1961 and 1968, reported by the Soviet researchers Kondratyev and Nikolsky. The later study also brings out another curious detail – the amount of heat received by the Earth from the Sun seems to be greatest when there is an intermediate number of sunspots, neither very many nor none at all.

Putting the numbers in, it seems that the greatest value of the solar parameter, about 1·94 calories per square centimetre per minute, occurs when the sunspot number is between 80 and 100, with a decrease in heat measured both for lower and higher sunspot numbers. This is immediately interesting on the time-scale we are looking at, since some solar maxima never reach a sunspot number of 100, some get to about this middle ground, and some reach much higher peaks. If the amount of heat reaching the Earth varies in this curious way with sunspot number, then the size of each sunspot peak is clearly going to be very important in determining the average amount of heat received by the Earth over each 11-year cycle – and therefore determining how temperatures on Earth change as the decades go by. Two related studies carried out at the US National Center for Atmospheric Research in the mid-1970s showed just how important this kind of variation is, in terms of the climatic changes we have experienced in the past few centuries.

The most striking 'coincidence' between the record of solar variability and the record of the changing climate is that just in the years and decades when the Little Ice Age was reaching its maximum intensity there were no sunspots at all on the Sun – no cycle of maxima and minima for about 70 years, ironically coinciding almost exactly with the reign of Louis XIV of France, the 'Sun King', from 1643 to 1715. Now, sunspots were only rediscovered by Galileo early in the seventeenth century, and it has been argued that rather than there being no sunspots at this time, nobody bothered to observe them. This has always seemed rather unlikely, since the discovery of the spots on the face of the Sun caused violent controversy in Galileo's time, and evidence for or against their existence must have been sought avidly in the century that followed. The most complete survey of the historical record of solar activity ever attempted was reported by Dr John Eddy in 1976, providing conclusive evidence that this suspicion is correct, and that for most of the second half of the seventeenth century there really were no sunspots to observe. Indeed, 'the total number of sunspots observed from 1645 to 1715 was less than what we see in a single active year under normal [*sic*] conditions'.

This prolonged absence of solar sunspot activity has become

known as the 'Maunder Minimum', after E. W. Maunder who drew attention to it almost a century ago. Another pioneer, on whose work Maunder's was based, has not even had his name immortalized in this way – but Eddy has now suggested that there may have been a previous extended minimum between 1460 and 1550, which might now be called the Spörer Minimum in recognition of the work of F. W. G. Spörer. It was Maunder, however, who quoted the comments of astronomers detecting the occasional rare spot in the late 1600s: 'In 1671, Cassini commented "... it is now about twenty years since astronomers have seen any considerable spots on the sun" '; Picard '... was pleased at the discovery of a sunspot since it was ten whole years since he had seen one'; and so on. Eddy points out that this was the time when Newton produced the reflecting telescope, and in which refractors with focal lengths of 2 to 4 metres and apertures of 5 to 10 cm – certainly comparable with the solar observing instruments of the eighteenth and nineteenth centuries – were being used to study the Sun. So the lack of observations of sunspots was neither due to lack of interest or poor equipment; it must have been due to a genuine lack of sunspots.

The second-hand evidence points the same way – the 'period between 1645 and 1715 was characterized by a marked absence of aurorae' ... 'far fewer were recorded than in either the 70 years preceding or following'. And the records come not only from Europe but from Japan, China and Korea; 'the oriental data (sunspots and aurorae) confirm that there were no intense periods of solar activity during the Maunder Minimum, and probably no "normal" maxima in the solar cycle'. Once again, however, the clincher seems to be the carbon-14 evidence, which not only confirms the reality of this long period of solar quiescence, but extends the record back into the Middle Ages, showing the quiet solar period of the Spörer Minimum from 1460 to 1550, itself preceded by a 'grand maximum' of solar activity in the twelfth and thirteenth centuries. Looking further back still, it becomes necessary to disentangle the changes caused by the Earth's changing magnetic field in order to interpret the direct solar activity influence on carbon-14 – but this goes beyond the thousand years or so of immediate climatic history that we are especially interested in, and already there is enough evidence to tie recent climatic changes firmly to the changes in the flickering solar furnace. The broad picture shows a grand maximum of solar activity coinciding with the Little Climatic Optimum, and the Spörer and Maunder Minima coinciding with the dips of the Little Ice Age. Coming closer to home, the warmth of the middle part of this century coincided with a period of increasing solar activity from one 11-year maximum to the next – a trend which has now (since 1959) reversed. Most important of all, as Eddy says:

> there is good evidence that within the last millennium the sun has been both considerably less active and considerably more active than we have seen it in the last 250 years . . . The reality of the Maunder Minimum and its implications of basic solar change may be but one more defeat in our long and losing battle to keep the sun perfect, or, if not perfect, constant, and if inconstant, regular. Why we think the sun should be any of these when other stars are not is more a question for social than for physical science.

Even while this chapter was being written, new evidence was published in support of the view that our Sun is not the constant, stable heat source of astronomical tradition. Just as remarkable as the new evidence is the sudden way in which the astronomical establishment can now be seen swinging into line with views that were regarded as eccentric only a few years – indeed, scarcely more than a few months – ago. In one single issue of the science journal *Nature*, dated 31 March 1977, we found a whole series of articles and scientific papers relating to this same theme. Dr David Hughes summarized much of the work of John Eddy, also outlined here, in a major feature headed 'The Inconstant Sun'; Eddy himself, writing with Dr Charles Smythe, reported new studies of solar changes in and after the Maunder Minimum; while, tucked away at the back of the journal short contributions from both Dr M. S. Muir, a South African, and Dr J. M. Colebrook, a British oceanographer, noted and discussed a 'very impressive correlation between sea-surface temperature and annual sunspot numbers' (Colebrook's words) for the North

Atlantic region. Suddenly, the concept of solar inconstancy is respectable; and with this new aura of respectability for the whole topic there must be real hope that the climatic forecasts which studies of past flickers in the solar furnace make possible will now also be taken seriously.

The past three centuries

The broad picture of the past millennium confirms our dependence on the changing effects produced by our imperfect, flickering Sun. When we look in more detail at the record of the past three centuries, we see just how close the links between solar activity and climate are, even on the small scale. Dr Stephen Schneider – like Eddy, from NCAR – and Dr Clifford Mass, of the University of Washington, went one step better than most theorists by including both the sunspot and volcanic dust effects in one computer model, looking at events since AD 1600. The changing transparency of the atmosphere produced by volcanic eruptions can be included in the calculations using Lamb's Dust Veil Index, and the changing influence of the Sun is included by using the relationship between the solar parameter and sunspot number that was found by Kondratyev and Nikolsky. Once again, it is important to remember here that we do not know – and it does not matter – whether the Sun's heat output changes, or whether some change in the 'colour' of the Sun's radiation affects the transparency of the ozone layer. Either way, the effect on radiation getting through to the ground is the same – the value of the solar parameter measured on Earth (or at the bottom of the stratosphere) is 2% lower for no sunspots than for sunspot numbers between 80 and 100, and back to the level corresponding to no sunspots for sunspot numbers of around 200.

Modern computer models are well up to the task of 'predicting' how the temperature of the Northern Hemisphere 'should' have varied since 1600 with the appropriate solar variations and DVI changes fed in, and the result is strikingly like the real temperature record of the past three centuries (Fig. 6.1). The model produces a clear 'Little Ice Age', a warming in the eighteenth century, a dip in the early nineteenth century and even the detailed warming of the early part of the twentieth century, followed by the cooling we are now experiencing. The obvious difference between the model and reality is that in the real world we do not see such big swings of temperature over the 11-year cycle – it seems that what matters is not the year-to-year changes in sunspot number, but the average level of activity of a whole 11-year cycle. That seems reasonable enough – after all, it takes some time for the atmospheric circulation and the whole climatic system to shift in line with changing outside influences, and this inbuilt delay must act to average out the more rapid flickers. After all, we don't see dramatic climatic changes overnight, even when the Sun is gone from view, because the 'memory' of the atmosphere and climatic system is so much longer than a few hours. It's the same sort of effect, but over a few years, that smooths out in the real world the sharp ripples in the model temperature curve of Fig. 6.1. The sharp rise in the model temperature after about 1968 is probably one of these spurious ripples.

Now, although we cannot predict the changing level of volcanic activity very easily, it does seem from this study that the main driving force behind temperature changes on the scale we have experienced since the Little Ice Age is the solar effect – even doubling the dust effect doesn't change the basic pattern of the model's 'predictions'. Having found that on a simple historical basis the worst climate we can reasonably expect would be a repetition of the Little Ice Age, we have also found that the flickers of the Sun can explain just that kind of event. We also found that these flickers seem to vary in line with some well-established cycles, and some rather less well-established ones; and, by no means least important, we found a plausible physical cause of these changes in the way the changing positions of the planets act to shake up a rather sickly, off-colour Sun. All this adds up to provide something near to a package for predicting climatic changes, by predicting future trends in these

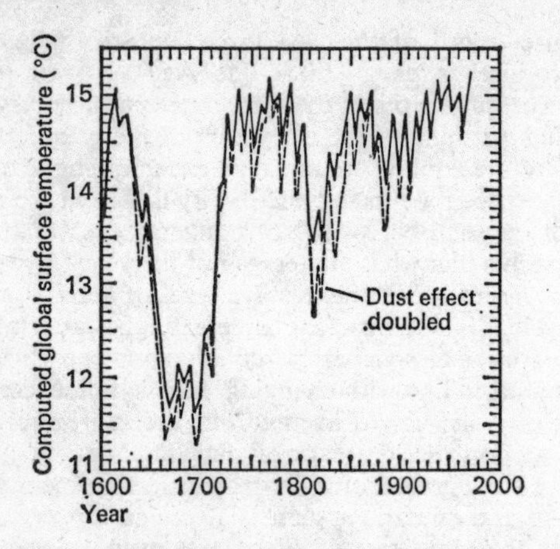

Figure 6.1 Surface temperature variations of the Earth since 1600 'predicted' by a computer model which takes account of changes in the solar parameter with changing sunspot numbers, and the changing Dust Veil Index caused by volcanic activity. The sunspot effect dominates, as can be seen by comparing with the variation for a model in which the expected Dust Veil Effect was arbitrarily doubled. Based on Figure 2 of the paper by Stephan Schneider and Clifford Mass, referred to in the text. (Compare with the actual temperature changes in the Northern Hemisphere, shown in Figure 2.1, p. 41.)

solar cycles. It would still take a brave man to actually make such a prediction in detail – but such a brave man exists, in the form of Professor Hurd C. Willett, of the Massachusetts Institute of Technology, who made just such a prediction in the *Technology Review* in 1976.

What the cycles foretell

Professor Willett has long believed in the reality of these solar cycles and their effects on climate, even before the definitive new work reported earlier in this book. Even in 1951 he forecast on

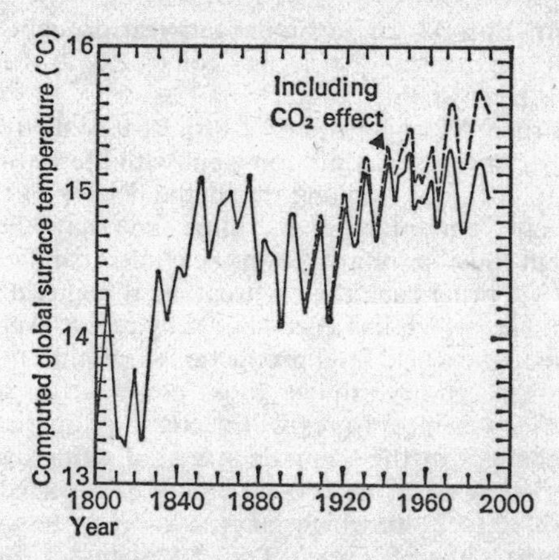

Figure 6.2 Global surface temperature computed by the climatic model, for the period since 1800. This includes a 'forecast' both with (. . .) and without (—) possible effects due to increased carbon dioxide (see Chapter 10). After Schneider and Mass.

this basis that temperatures would fall over the next 15 years, a forecast which fell on deaf ears at a time when most meteorologists were worried about the warming trend of the previous decades, that had only just been noticed! The value of the new physical understanding of the solar–terrestrial links is that we can now take these predictions without the pinch of salt that used to be deemed necessary, and it seems only fair to report Professor Willett's latest long-term forecast in some detail – it is, after all, the best long-range forecast now available.

According to this forecast, based on the various solar cycles, the next 25 years will be significantly cooler in all latitudes than the years of the mid-1960s. The decline may start immediately, says Professor Willett, with minimum temperatures reached in the 1980s, or there may be a short delay before the decline sets in, so that minimum temperatures are reached in the 1990s. At higher middle latitudes, and in the sub-tropics, the next two decades will be dry, with particularly severe droughts in Canada and northern Europe, and a similar ten-year drought occurring in Asia and sub-tropical Africa may coincide with the northern droughts, or begin slightly later.

At the turn of the new century, Professor Willett expects a sudden return to warmer conditions, but with the warmth of the years 2000 to 2010 not reaching the idyllic heights of the period 1930–60. Wetter conditions at the same time may alleviate the droughts, but then a return towards cooler conditions after about 2010 will bring back the dry weather at higher middle and tropical latitudes which has been shown by recent events to put severe strain on the world food production system. For the rest of the twenty-first century, things look pretty grim, and, says Professor Willett 'temperatures will decrease dramatically' with associated changes in the 'climatic stress' affecting agriculture. The next period as warm as 1930–60 cannot be expected until the years from 2110 to 2140, which takes us more or less up to the end of the next 180-year 'cycle' of solar variation.

Predictions beyond 180 years ahead must necessarily be more vague, but there are predictions based solely on the solar–climatic effects. These ignore changing volcanic activity, of course, which must make them a little suspect – and they also ignore some longer, slower astronomical effects which may be the basic cause of full Ice Ages, and which will be discussed in the next chapter. But on a scale of up to 10,000 years or so the solar–climate theory is as good as any for predicting future climate – and better than most. Professor Willett places particular emphasis on two cycles, one of 720 years (which, of course, is 4×180 years) which is suggested by the record of changing solar activity (carbon-14, aurorae and so on), and the other roughly 10,000 to 12,000 years long, which is indicated by the 'second-hand' climatic record, but cannot be related to solar activity changes since there is insufficient evidence of how the Sun may vary on such a timescale. The most severe period of climatic stress on record, as Professor Willett points out, was between 1370 and 1390, when all of northern Europe was ravaged by:

> strong blizzards and extreme cold in winter (including record storm flooding of the Dutch lowlands) and heat and

> drought in summer. The consequences in famine and plague reduced the population of the British Isles by two-thirds. Old Chinese records indicate that *this was a period of sunspot activity such as has never been seen since.*

The italics are mine – for this fits in strikingly with the idea that very few *or very many* sunspots produce the same pattern of erratic variation in the Earth's climate. This period was the prelude to the Little Ice Age, since when sunspot activity has shown an upward trend over three centuries or so. With this clear-cut event as a 'benchmark' to establish our position today within the cycles, and using the 720-year and 10,000-year cycles alone, the prediction is that the climatic stress expected in the period following the warming in the twenty-second century will match the events of 1370 to 1400 and may be followed by another Little Ice Age, from 2200 to 2550, probably 'more severe than that from 1500 to 1850'. And that is as far as the solar–climate theory can take us. As we saw from the work of Schneider and Mass, the theory adequately explains the most recent Little Ice Age, so it is sensible to use it to predict the next Little Ice Age. But there doesn't seem to be any need to invoke even bigger solar fluctuations in order to explain full Ice Ages, because we have available another theory to predict the next Ice Age itself, a few thousand, rather than a few hundred, years ahead.

In the very long term, indeed, we expect the Sun to return to its more normal state, with nuclear 'burning' starting up again and raising the temperature inside by about 10%. When that happens, the present cycle of Ice Ages will be over, and the tens of millions of years of warmth and plenty that are the natural state of the Earth will return – until something sends the Sun off the boil again for another temporary cool spell. The very small ripples on climate (up to and including Little Ice Ages) are the obvious symptom of the Sun not being normal today, and running slightly cooler overall than our computer models suggest it should. These little ripples do not seem to produce Ice Ages, but because the climate has been pushed that much closer towards an Ice Age only a small outside effect is needed to tip the balance completely. That extra influence seems to come from another astronomical effect – a change not in either the quality or quantity of radiation being produced by the Sun, but a change in the position of the Earth within that sea of heat and light. Because the Earth's orbit is not quite constant, but is slowly changing, we do not always receive the same pattern of summers and winters – and that, it now seems, is the basis of the known pattern of full Ice Ages and their variations.

Chapter Seven

The next Ice Age isn't due just yet

It was in 1976 that a combination of several pieces of evidence fell into place to confirm that climatologists now understand the basic cause of the rhythmic ebb and flow of the glaciers that has characterized the more recent series of Ice Ages here on Earth. Probably, the same processes also affected the changing climate during the previous ice epochs, hundreds of millions of years ago. But the greatest significance of the new breakthrough in understanding of Ice Ages is that it provides a firm prediction of how the natural course of events will develop – if undisturbed by man's activities – over the next few thousand, or tens of thousand, years. We can now say unequivocally that the warmest period of the present 'interglacial' is over, and that from here on we can expect a cooling off until within about 10,000 years the world will be in the grip of another full Ice Age.

Depending on your point of view that is either worrying or reassuring. The prophets of doom, comparing this prospect with the very long history of the Earth, point out that, by those standards, the next Ice Age is 'just around the corner'. On the other hand, optimists such as myself argue that the evidence shows that we do not have to worry about the ice for a couple of thousand years yet – and if we survive that long, we should be able to do something about preventing the spread of ice. Indeed, as we shall see in Chapter Ten, man *already* has the ability to stop the next Ice Age happening. The astronomical theory of Ice Ages, which makes these predictions possible, is generally called the 'Milankovich Model', after Milutin Milankovich, of Yugoslavia, who spelled out details of the idea some 40 years ago. The basis of the idea is much older, though, and among others who supported it even before Milankovich was Alfred Wegener, better known to us today as one of the fathers of the concept of continental drift. Like that concept, the Milankovich Model of Ice Ages has survived for decades before at last being proved basically correct.

Standard textbooks, such as Arthur Holmes's *Principles of Physical Geology*, have long acknowledged the existence of the model, and spelled out the details of the basic mechanism. Three separate cyclic changes in the Earth's movements through space combine to produce the overall changes in the solar radiation falling on the Earth, and are the key to the theory. The longest of these is a cycle of between 90,000 and 100,000 years, during which the shape of the Earth's orbit around the Sun stretches from almost circular to more elliptical, and back again. When the orbit is nearly circular, there is a more even spread of solar heating over the year, taking the Earth as a whole; when the orbit is elliptical, we are closer to the Sun at some times than at others, and this effect can act to increase the contrast between seasons, even though the total heat received by the whole Earth over an entire year may stay the same.

The second effect is a cycle some 40,000 years long during which the tilt of the spinning Earth changes, with the Earth 'nodding' up and down relative to the imaginary line joining the centre of the Earth and the centre of the Sun – known technically as a change in 'the obliquity of the ecliptic', this effect directly changes the contrast between seasons (Fig. 7.1). When the tilt is more pronounced there are strong seasonal changes, and when the Earth is nearly 'upright' then there is less difference between summer and winter. Finally, the gravitational pull of the Sun and Moon on the bulging equatorial regions of our planet produces a wobble like that of a spinning top – but with a period of between 20,000 and 25,000 years. This is the 'precession of the equinoxes'.

These effects combine to produce changes in the amount of heat arriving at different latitudes at different times of the year – but they do *not* change the total amount of heat received from the Sun by the whole planet over a whole year. The details of the Milankovich Model are spelled out in my book *Forecasts, Famines and Freezes*, and there is no need to go into them here. But it is very easy to see in general terms how this kind of change

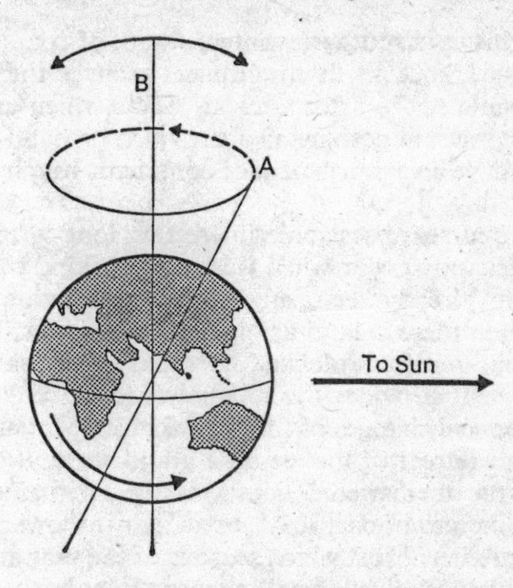

Figure 7.1 The 'nodding' Earth. The Earth spins every 24 hours about axis A. Axis A wobbles around axis B, roughly every 21,000 years. Axis B 'nods' up and down relative to the Sun, roughly over 40,000 years.

in seasonal heat could encourage ice to spread – given the present positions of the continents. It is clear that cool summers in the Northern Hemisphere might allow the snow which falls in winter on the land surrounding the polar sea to persist through the summers. Once some snow and ice fields become established in this way, we can imagine that, by reflecting away a good part of the weak summer heat, they will encourage the rapid spread of glaciation through a feedback process.

On the other hand, the conditions we need to produce a spread of ice in the Southern Hemisphere are just the opposite. What is important there is to have very severe cold winters, in order to freeze more ice from the sea – snowfall alone is no use, since this has no free land to fall on, and so a cool summer won't help the ice to spread as much as a severe winter will. What we need for global Ice Age conditions, then, are cool northern summers plus cold southern winters – and, of course, the two go hand in hand! Since northern summers occur at the same time as southern winters, the astronomical effects which are needed to produce Ice Ages in both hemispheres also go hand in hand – as long as we have an arrangement of continents roughly like that of the present day.

This, of course, raises some interesting long-term questions – can Ice Ages only occur when we have this kind of geography? As we have already seen, any kind of glaciation only seems possible when there is land at high latitudes; now, it looks as if the combination of land-locked northern polar sea and a south polar continent provides the uniquely necessary conditions for the astronomical changes of the Milankovich Model to cause the advance and retreat of the ice on a grand scale. But for years – decades – the unanswered questions have remained: Has the fluctuating pattern of past Ice Ages actually followed the changes the model predicts? Just which seasons of the year are the critical ones, in which relatively small changes in solar heating trigger spread of ice and snow cover? And are the changes in solar heat (insolation) really enough to explain the spread of ice or to melt the glaciers at the end of an Ice Age? With many climatologists eagerly investigating the problem, it was only a matter of time before the answers came – but it was something of a coincidence that all three questions were answered in the same year, 1976.

Part of the reason for this 'coincidence' was the improvement in computer modelling techniques that took place in the early 1970s. One example of the power of the new computer models came when Professor Johannes Weertman, of Northwestern University in Illinois, used the technique in an analysis of how ice sheets grow under the influence of small changes in heat radiation. He found just the kind of changes needed to explain Ice Ages by the Milankovich Model, confirming that this could be the true cause of Ice Ages. But as he said, the evidence did not on its own settle the issue once and for all – the results demonstrated 'that the Milankovitch radiation variations are potent enough to be the prime cause of ice ages, although [they] by no means prove that the Milankovitch ice age theory is correct'. Various other attempts have also been made to produce computer models of how ice sheets vary, and they generally produce the same kind of results, showing that the Milankovich variations *could* cause Ice Ages, but not *proving* that they do.

Another aspect of the modelling approach is to use our present understanding of the way the reflectivity of the Earth changes, and of the changing amount of heat stored up in the oceans, combined with the present geography, to work out a 'global heat budget' and find out which past epochs should have been so deficient in heat – or overdrawn on the budget! – that ice must spread. Dr David Adam, of Menlo Park in California, is one of the modellers who have tried this approach recently, and he came up with one curious, but perhaps important, result. According to Adam's calculations, Northern Hemisphere ice sheets could form without producing a dramatic change in the climate of the Southern Hemisphere, because the land-based ice sheets would have little effect on overall ocean circulation. But changes in the southern ice must, as we saw before, affect the whole globe by changing the whole pattern of ocean currents. Adam suggests that perhaps two mechanisms act to cause Ice Ages – the Milankovich effect with its more or less regular, periodic changes, together with the less predictable situation of Antarctic ice surges. These surges might themselves be triggered by changes in snowfall and seasonal temperatures produced by the Earth's wobbles through space, perhaps, but we should remember that not all Ice Ages need be caused by the Milankovich mechanism. That prediction of a new Ice Age within about 10,000 years assumes not only that man does not interfere with nature, but also that there is no surge of the Antarctic ice in the intervening time. Perhaps the prophets of doom would be better occupied watching the southern ice cap than studying the Earth's movement through space.

One of the most succinct summaries of just where this kind of modelling leaves the Milankovich Model was provided by Drs Max Suarez and Isaac Held, of Princeton University, in a short paper published in *Nature* in 1976. Their model is relatively sophisticated, and can be used to predict the temperatures of two layers of the atmosphere (representing the upper and lower halves of the troposphere), the surface temperatures over land

and ocean, the depth of snow over land, and the thickness of sea ice, all calculated for different latitudes and different seasons as required. By changing the incoming solar radiation in line with the Milankovich Model, the Princeton team was able to 'predict' how the climate of the past 150,000 years should have varied. This study has one particular value, because the nature of the model means that seasonal changes are specifically picked out, showing that, for example, changes in the warmth of Northern Hemisphere summer are particularly important in affecting the overall climate, just as we suspected. Overall, although the 'predictions' do not match the actual record of the past 150,000 years precisely, the agreement is close enough for Suarez and Held to conclude 'that a substantial portion of climatic variability on these time scales can be understood as the equilibrium response to perturbations in the orbital parameters' – in other words, a good part of the climatic changes are caused by the Milankovich mechanism. Calculations made by Dr George Kukla, of the Lamont-Doherty Geological Observatory in New York State, pinned down the critical season for the insolation changes even more precisely, as the Northern Hemisphere late summer/early autumn, with the interiors of North America and Eurasia seen as the key geographical locations for the Milankovich influence on snow and ice cover to be felt. But we were still left with the two key questions to answer – did the record of past Ice Ages show really convincing evidence of the Milankovich cycles, and was there enough energy involved in these fluctuations to do the trick?

Pinning down the periods

The breakthrough in determining just how temperatures and ice sheets have fluctuated over the past few hundred thousand years may have been published only in 1976, but it was certainly a long time in the making. It's necessary to have such long-range, accurate information if you want to test for the reality of a cyclic change as much as 90,000 to 100,000 years long, just as we need several hundred years of evidence to test for climatic effects with periods of a few tens of years; and until recently such long-range information on climate that we had was very sketchy, incomplete and, we now know, downright inaccurate.

On the old picture of Ice Ages, geologists reckoned that there had been four or five periods of greatly increased Northern Hemisphere ice cover in what, to a geologist, could be regarded as the 'recent' past. But new evidence, built up over the past 20 years or so, shows that in fact there have so far been something like a score of these glacials, separated from each other only by rather short warm periods, or interglacials, during the history of the current ice epoch.

It's hardly surprising, then, that Milankovich himself and other supporters of his theories could not make the model fit the 'evidence' in the 1930s, 1940s or 1950s. The Milankovich Model 'predicted' many more Ice Ages, much more closely spaced through the recent history of the Earth, than the geologists had evidence for. When that was the best evidence available, the Milankovich Model had to be left out in the cold. But as new techniques have provided ever better evidence of how the Earth's climate changed over the past few hundred thousand years, the picture built up from the evidence has changed repeatedly; every change has brought the evidence more nearly in line with the Milankovich Model, until now there remains no doubt at all that the periodic ebb and flow of the great glaciers does indeed include cyclic changes in line with the rhythms of the model.

This is not the place to describe at length the revolution in geophysics that included a complete revision of the timescales attributed to Ice Ages,* but the key technique which led ultimately to the breakthrough in establishing the reality of the Milankovich cycles is worth spelling out in a little detail.

The technique depends upon two factors – a better supply of raw material from which past climates can be reconstructed, and better methods for carrying out the reconstruction. The first part, the raw material, is now provided by research ships that cruise the oceans of the world, retrieving cores of mud and other debris from different parts of the sea bed. As the years go by,

* See my book *Our Changing Planet.*

sediments are deposited on the sea bed, building up this ooze in layers which correspond to the passing years, decades and centuries. Old sediments, deposited perhaps hundreds of thousands of years ago, lie deep in the mud, while new sediments, deposited only yesterday, are at the top – and in between lie the sediments from all the years in between. These sediments cannot be interpreted like tree rings – there are no obvious thick and thin layers to interpret in climatic terms. But there is evidence of climatic change there to be revealed by modern analytical techniques.

In particular, the techniques used with great success in analysing ice cores can be applied to these ocean sediments. That depends upon measuring the proportion of the heavy isotope oxygen-18 present, because this proportion is affected by temperature. For the ice cores, the technique is relatively straightforward, since after all every molecule of water or ice contains its oxygen, which in principle at least can be analysed. But where is the oxygen of the past locked up in ocean sediments for us to study? The answer is, in the shells of small marine creatures, the fossil remains of the animals known as foraminifera. These shells are basically made of chalk, chalk which contains oxygen atoms locked up in compounds with other elements, oxygen atoms plucked from the air of the Earth at the time the creatures were alive, and preserved where we can now measure the proportion of heavy oxygen present and use the measurements to work out a record of temperature changes covering hundreds of thousands of years.

There is also a bonus here for the workers trying to pin down this climatic timescale. One of the other principle components of the chalky remains of the foraminifera is carbon, and that carbon too is representative of conditions on Earth at the time the animal was alive. The radiocarbon (carbon-14) dating technique can be applied to these shells, at least for the more recent part of the cores (carbon dating is less effective for longer timescales, but other radioactive tracers can be used to lengthen the 'calendar').

As long ago as 1955, Dr Cesare Emiliani, working in Chicago, established from the oxygen-18 fluctuations in the fossil remains from cores drilled in the Caribbean, that there had been many more abrupt changes in climate in the past few hundred thousand years than could be fitted into the then conventional picture of Ice Ages. Hardly surprisingly, geologists did not welcome this evidence, and it was not until the late 1960s that, with so many of their cherished beliefs overturned by the new understanding of plate tectonics (see Chapter Three) and continental drift, a revolution in the understanding of past climates proved possible. Improved techniques on land also showed many more Ice Ages than the handful that had been accepted, and the clinching evidence came when improved dating techniques showed that the evidence of Ice Ages from the land record did indeed match up in time with the evidence from the oceans.

By the 1970s, the time was right for a new investigation of the Milankovich Model, and techniques had advanced to provide even more evidence about the pattern of past Ice Ages. As well as the direct measurements of oxygen-18 in the sediments, the number and nature of the fossil shells themselves can tell us about the climate at the time their inhabitants were alive. Just as we do not find polar bears living in tropical regions, or humming birds in the Arctic, so the different species of foraminifera have their own preferred climates. One family of these tiny animals may prefer colder water, while another prefers slightly warmer water – and by counting the number of each animal in different sediment layers it is possible to get another handle on the changing climate. And another new technique allows the experts to work out from the fossil foraminiferal remains how much ice there was over the globe at different times in the past.

As well as the direct effect of temperature on the amount of oxygen-18 getting into the air and being absorbed by living creatures, the amount of ice cover also plays a part. The molecules of water that contain oxygen-18 instead of the lighter oxygen-16 do not move so speedily as the lighter molecules, and as a result it is easier for them to stick on to a growing ice sheet and freeze. So the ice sheets end up with a high proportion of oxygen-18, leaving relatively little to be absorbed by sea creatures (or anything else). At first sight, this looks like a problem. After all, at face value a high proportion of oxygen-18 in the fossils should indicate cooler conditions – but at the same time if the ice sheet effect has been at work then we might expect *less* oxygen-18 in the fossils from Ice Ages. But the answer is stunningly simple. First, the changing oxygen ratios in creatures that are known to have lived in the deep ocean are measured. Because these creatures are hardly affected by temperature changes at the surface at all, the varying proportion of oxygen-18 revealed can be used to calculate how the ice cover of the globe has changed. Then, with this effect allowed for, oxygen ratios from different species, surface dwellers that are strongly influenced by global temperature changes, can be interpreted as a direct measure of changing temperature – a global thermometer.

With thousands of sea-floor cores now available, and two decades of accumulated expertise in interpreting them, Drs Jim Hays, John Imbrie and Nicholas Shackleton put together the most definitive study of the cycles of ice advance and retreat over recent millennia yet available. The results, published in *Science* late in 1976, were based on detailed study of two selected cores, one of which provided a good continuous record over the past 300,000 years while the other, although spoilt at the top, gave a good record for the period between 100,000 and 450,000 years ago. With the overlap in the middle providing a check of one core against the other, the net result was a continuous record of temperature and ice cover changes going back for almost half a million years – sufficient time to test for the presence of even the 90,000- to 100,000-year cycle.

The various different climatic tests were applied to samples taken from these cores at intervals of 10 cm throughout their length, which combined to give a 15-metre span covering the past 450,000 years – just 150 sample 'dates' in all, providing a spacing of about 3000 years between each date, which is sufficient to pick up any cycles as short as 20,000 years. So the three Milankovich cycles were nicely bracketed by the evidence, and all that remained was to test for the existence of the cycles, using the statistical analysis techniques which pick out the 'spectrum' of periods present in any 'signal' that contains many interacting cycles, and some random effects, or 'noise'. Different tests on the different climatic indicators produce slight differences in the detailed numbers that come out of this analysis, but Dr Hays and his colleagues found close agreement between the results, confirming the reality of three significant cycles:

> There can be no doubt that a spectral peak centred near a 100,000-year cycle is a major feature of the climatic record . . . dominant cycles . . . range from 42,000 to 43,000 years [and] three peaks . . . correspond to cycles 24,000-years long.

Of these effects, the shorter period influences are more dramatic, while the roughly 100,000-year-long cycle of 'stretch' in the Earth's orbit acts to modulate the influence of the shorter cycles, with the coldest periods of Ice Ages coinciding with times when the Earth's orbit is more nearly circular, so that there is no counter-effect from this cycle to offset the deepest dips in the other cycles. The roughly 23,000-year-long cycle seems to have had a strong effect in recent millennia, producing changes in climate almost immediately as the astronomical influence changes – and the statistics also reveal a less important 19,000-year cycle, which is explained by the latest astronomical calculations as a secondary effect of the precession of the equinoxes cycle.

In some ways, the 42,000-year period is the most intriguing, since although there is a clear impact of the cycle on climate over a period corresponding to ten of these 'rolls' of the Earth, and including four major Ice Ages, the climatic cycles seem to follow about 8000 years after the astronomical cycles. This can be explained in terms of the delays built into the climatic system of the Earth – the effect, perhaps, of the build-up and subsequent surge of the Antarctic ice sheet, or of other, more subtle, adjustments of the air, sea and ice to changing outside influences. Hays, Imbrie and Shackleton have no doubt, however, about the conclusions to be drawn from their study:

> It is concluded that changes in the earth's orbital geometry are the fundamental cause of Quaternary ice ages.
>
> A model of future climate based on the observed orbital–climate relationships, but ignoring anthropogenic

> effects, predicts that the long-term trend over the next several thousand years is towards extensive Northern Hemisphere glaciation.

This evidence was alone sufficient to convince many people, especially the geophysicists and astronomers, of the accuracy of the Milankovich Model. The latest calculations of the detailed changes in the Earth's orbit and inclination, carried out by Professor André Berger in Belgium, and Professor A. D. Vernekar in the US, lend further support to the model, with the refined calculations (improved with the aid of electronic computers) giving even better agreement with the climatic record than the more approximate calculations of Milankovich himself and others through the decades. But all this has one snag – it depends on rather sophisticated techniques, first in analysing the core samples, then in pulling out the periodic cycles from the statistical hat – and even the astronomical calculations of the expected cycles now depend, it seems, on the modern high-speed electronic computer. This does nothing to invalidate the results, of course, but it would be pleasant to have some simple, easy-to-understand test of the reality of the Milankovich Model, which answered the question 'just how do the small changes in seasonal heat allow ice to build up so dramatically, or melt the glaciers so quickly?' Happily, this last piece of the Milankovich puzzle was fitted into place, also in 1976, by Professor B. J. Mason, the Director-General of the UK Meteorological Office. Mason's investigation of the Milankovich Model is a classic of simplicity, and so convincing that he persuaded even himself that the model must be right, after having previously been one of the most outspoken opponents of these ideas! Since you don't need a degree in mathematics to understand the calculations either, Mason's treatment of the problem surely deserves special mention, marking the moment when the Milankovich Model finally came in from the cold, to be welcomed not just by specialist geophysicists and astronomers, but by the meteorological establishment, in a form intelligible to everybody.

Where the energy comes from – and goes to

The steady build towards establishing the Milankovich Model was crowned by a study of the heat energy balances involved – the Earth's energy budget – so simple that in retrospect it seems astonishing that no one had thought of it before. It does depend on a good understanding of how the amount of ice covering the Northern Hemisphere has varied over the past few hundred thousand years, but once this vital piece of evidence is available the energy budget approach cuts to the heart of the problem with classic simplicity, and depends not on any sophisticated statistical analysis involving expensive, high-speed electronic computers but on some simple sums that can be literally carried out on 'the back of an envelope' – the physicists' usually mythical criterion of absolute simplicity. In order to melt ice, heat must be supplied to overcome the 'latent heat of fusion' at a rate of 80 calories for every gramme of ice. In other words, in order to turn a given amount of ice at 0°C into water at 0°C, the same amount of heat is necessary as would be required to warm up the same amount of water from 0°C to 80°C. And when you are talking about melting glaciers, that adds up to a lot of heat. The need for this heat is that water is a more energetic form of matter than solid ice – and the gaseous form of water is more energetic still, with a latent heat of evaporation needed to turn water into vapour several times greater than the latent heat of fusion. When water vapour in the air, at a temperature of 0°C, condenses into water each gramme liberates almost 600 calories of heat – enough to melt several grammes of ice; and if the water vapour goes the whole way to solid, and falls as snow, the total latent heat released is about 675 calories for each gramme. This heat must go into warming the surrounding air, and spreads throughout the globe as a distinct influence on the overall temperature. Equally, the need for heat to be absorbed in melting snow or ice helps to keep regions of the globe covered by winter snow cool well into the spring and early summer. Only when the snow has been melted can the warmth of the summer Sun be used to heat up the land

and air near the ground. Each year, we can see the effects of latent heat processes as the seasons progress, bringing us a cycle rather like an Ice Age in miniature. We know that ice cannot build up immediately the temperature drops below 0°C, but only when, and if, the total deficit on the heat budget is sufficient to overcome the latent heat barrier; equally, snow or ice cannot disappear as soon as temperatures rise above 0°C, but only slowly as the surplus on the heat budget builds up to push past the barrier in the other direction. It is no good melting one gramme of ice if the heat absorbed is sufficient to freeze two other grammes, or freezing one gramme of water vapour if this liberates enough heat to melt several grammes of ice. The deficit or surplus in the budget must be sufficient to do the job *and* to absorb the corresponding latent heat fluctuations without taking the overall temperature past the critical 0°C mark.

So, just how do the changes in heat from the Sun – changes in insolation – over recent Earth history add up, compared with the amounts of latent heat that we know were involved in the spread and retreat of the glaciers of the most recent Ice Age? Professor Mason has made the necessary calculations, using both Milankovich's original estimates of the insolation changes, and the 1972 refinement of these estimates made by Professor Vernekar. Between 83,000 and 18,000 years ago, during the glacial period, there was an overall deficit in insolation, compared with the present-day situation, of no less than $4{\cdot}5 \times 10^{25}$ calories, equivalent to about 1000 calories for each gramme of ice that we know, from the geological record, was formed over that period to produce the great ice sheets covering the Northern Hemisphere. Vernekar's figures suggest a slightly smaller deficiency in the heat budget, $2{\cdot}5 \times 10^{25}$ calories overall, corresponding to about 550 calories for every gramme of ice formed. Either way, however, the simple calculations give impressively close agreement to the figure of about 675 calories which we know to be associated with the production of one gramme of snow at 0°C from one gramme of water vapour at 0°C.

From about 18,000 years ago to the present, all this accumulated ice has melted, except for a small residue near the pole. During this time, there has been an overall excess of insolation at high latitudes – a surplus on the heat budget – of between $4{\cdot}2 \times 10^{24}$ and $1{\cdot}0 \times 10^{24}$ calories. And the amount of heat needed to melt the volume of ice which we know did melt over the past 18,000 years is simply calculated as $3{\cdot}2 \times 10^{24}$ calories! Even though the calculations are so simple and approximate, the results are far too good to be explained by coincidence. We are dealing with mind-boggling, literally astronomical numbers, 10 followed by no less than 24 zeros (one million, million, million, million), and any 'coincidence' which stretches in this way across 24 'orders of magnitude' to provide such precise agreement cannot be dismissed lightly.

The way the Milankovich mechanism works can be seen, with hindsight to aid us, in Fig. 7.2, which is based on a figure prepared by Professor Mason. The change in ice volume is worked out from the known changes in sea level over the past 160,000 years, using the fairly obvious relationship that when more water is locked up in glaciers on land the sea level goes down. The variation in the insolation, or heat budget, is worked out both from Milankovich's figures and those of Vernekar, and plotted out for comparison with the ice-cover variations – Mason restricts the calculations to latitudes above 45°N where ice-cover variations actually take place, but this is an arbitrary 'cut-off' latitude, and it may be that this approximation is the reason why his calculations are not as yet quite perfect. When there is an excess of insolation, of course, we expect ice to melt, and when there is a deficit – when the curve falls below the central line in the diagram – we expect ice to build up. This is clearly what happened – but, equally clearly, there is more than just the Milankovich mechanism at work here.

About 140,000 years ago, increasing insolation brought about the rapid end of the most recent Ice Age but one. The next dip in the insolation curve did not bring about an immediate return of the ice, however, but only a small increase that was rapidly halted as the insolation swung back into a surplus. Some 83,000 years ago, another switch into deficit on the heat budget brought a strong build-up of ice, which was temporarily halted by the warmer period close to 50,000 years ago, and only built up to the maximum of the latest Ice Age when the insolation curve was

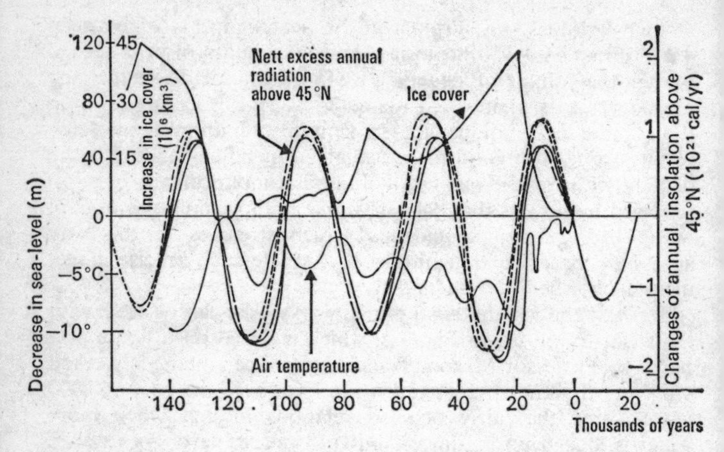

Figure 7.2 Changes in the global ice cover during the past 150,000 display a striking correlation with the curves for insolation of latitude 45°N computed by Milankovich (solid curve) 1930s and, more recently, by A. D. Vernekar (dotted). (Redrawn from Mason's paper in the *Quarterly of the Royal Metereological Society*, **102**, p. 4, 73, 1976.)

again in deficit just over 20,000 years ago. The past 18,000 years, with surplus on the heat budget, have seen the end of this phase of glaciation.

Looking at Fig. 7.1, it is easy to see that the Milankovich effect produces a very important cyclic ripple on top of other, longer-term fluctuations in climate. We know that there are very long-term influences at work, and also the likely erratic effects of volcanic dust thrown into the atmosphere. Because of this, we cannot use the clear cycles of the Milankovich Model to predict exactly what will happen in the next few thousand years – but this is the best model we have for making any such prediction. In the very long term, we are sure that Ice Ages can only occur when the geographical arrangement of the continents is suitable, as it is today; we suspect that even then there must be some

period expected for the years 2110–40 will not, after all, match the equable decades from 1930–60. Even the cooling Willett predicts for the twenty-first century, starting in less than forty years from now, could now be expected to be even more severe than he suggests. All these forecasts are immediately relevant to our present global society which should, on any sane picture, be making at least tentative plans for the next one hundred or two hundred years. The prospect of a return to a full Ice Age is grim, but remote on any human timescale. But as a backdrop to a picture of more immediate climatic deterioration, with each recovery failing to reach the previous peak, and each trough a little deeper than the one before, this prospect is one which should be in the minds of everyone concerned about the human race and its future as the owner-occupier of one rather small planet.

Chapter Eight

The seismic connection

Before we look in detail at the future of mankind under the threat of a major climatic change, there is one related topic which must be discussed. As well as the wave of freak weather conditions that affected many parts of the world in 1976, the year was unusual for the large number of damaging earthquakes that occurred, with extensive loss of life. Many people have asked whether the two might be connected – do the same root causes affect both weather and earthquake (or seismic) activity? The answer to this question can now clearly be seen to be 'yes', at least in part, and a look at this seismic connection can help us to understand just how much our small planet is at the mercy of the buffeting gusts of the solar wind, the wind that shakes the world.

Indeed, my own interest in the workings of the atmosphere and the way the Sun affects atmospheric circulation grew up just because I was involved in a study of some earthquake phenomena – a study reported in more detail in the book *The Jupiter Effect.* In that work which I carried out with NASA astrophysicist Dr Stephen Plagemann, it became clear that under appropriate circumstances some earthquake-prone regions of the Earth's crust can be shaken into seismic activity – causing both earthquakes and volcanoes – by the buffeting of the solar wind, acting through its influence on the atmosphere. Now, although earthquakes are very serious to anyone caught up in them, seismic activity only affects a few well-defined regions of the globe, danger areas that can be picked out using the modern theory of plate tectonics and continental drift. On the other hand, even a small shift in atmospheric circulation and weather patterns can have serious consequences for everyone on Earth. Because changes in climate clearly pose the more serious threat, world-wide, this is the area in which my own recent work has been concentrated. But the seismic connection is still there, and still something to be taken very seriously indeed if you are unfortunate enough to live in Los Angeles, China, Turkey or the other seismic centres of activity around the globe.

One of the key discoveries in this connection is the importance of many kinds of 'trigger effect' in setting off earthquakes. As great blocks of the Earth's crust drift about on the more fluid interior material of our planet, strain builds up as the edges of the blocks grind together, like pieces of a jigsaw puzzle being stirred about. When enough strain builds up, tension is released locally in an earthquake, or series of earthquakes, either great or small depending on local conditions. Clearly, when *almost* enough strain has built up for tension to be released it only needs a little extra push to send things over the top, in a kind of 'last straw' effect. Several trigger processes are known to operate in nature, including such apparently feeble effects as a change in the barometric pressure. Ronald and Barbara Oriti, of California's Griffith Observatory, summarized some of the other findings on earthquake triggers in articles in the *Griffith Observer*, late in 1976. It turns out that earthquakes are slightly more frequent at new or full Moon (when the tidal pull of Sun and Moon is exactly aligned), at times when the Moon is either nearest or furthest away from the Earth, and when the Moon is high in the sky rather than low on the horizon.

These tidal effects can trigger earthquakes which involve vertical movement of blocks of the Earth's crust, in particular, for which the gravitational pull effect operates in the same direction as the natural tendency for ground movement. Hardly surprisingly, since the pull exerted by the Earth on the Moon is twenty times greater than that exerted by the Moon on the Earth, many 'moonquakes' seem to be triggered by the corresponding tidal effects from the Earth, revealed by the four seismic stations left on the Moon by Apollo astronauts.

More distant cosmic influences may also affect earthquake activity. Drs Dror Sadeh and Meir Meidav, of the University of Tel-Aviv, reported as long ago as 1973 that there is a statistically significant tendency for an increase in the number of shocks in sensitive regions of seismic activity at the times when a certain region of the sky, in the direction of the centre of our Milky Way Galaxy, rises and sets. The chance of such an effect occurring through random effects is, they say, one in a million – but the explanation of the phenomenon remains a mystery. Could it be linked with bursts of gravitational radiation from collapsing black holes away towards the centre of our Galaxy? Or is there a more prosaic explanation closer to home? No one knows; but we do know from evidence such as this that the Earth is indeed sensitive to the delicate interplay of cosmic forces.

Coming closer to home, and to more practical aspects of this seismic connection, one of the massive earthquakes that were a feature of 1976, the major disturbance in eastern Turkey late in November, was actually predicted in advance by the Soviet scientist Professor A. I. Yelkin, using just the relationships between the Moon and earthquakes outlined above. Some idea of the controversial nature of this whole area of research can be gained, however, from noticing that while *Izvestia* reported Professor Yelkin's work in glowing terms, *Pravda* published an article dismissing the ideas in no uncertain terms as 'myths'.

No one, however, seems to be prepared to dismiss the very well-documented evidence of links between meteorological variations and small earthquakes, the so-called 'meteorological microseisms'. Dr Alois Zátopek, of Charles University in Prague, is one of the leading researchers in this field of study, and through the 1970s has provided geophysicists with a mass of evidence relating these phenomena to changes in the flow of air around the great circumpolar vortex. But something, of course, must cause the atmospheric circulation to change in the first place, and here Zátopek's discovery that the microseisms (very small earthquakes) also correlate closely with 'variations of the intensity of cosmic rays and frequency of occurrence of solar flares, associated with type IV radio emissions [from the Sun]', as well as being 'roughly indicated in Wolf's sunspot numbers' brings us right back into the mainstream of the work we have encountered so far in our look at the physical links between the Sun and the Earth.

There is, indeed, ample evidence of the way in which the

effects of variations in the solar wind during the Sun's cycle of activity affect the whole Earth, producing not just short-term effects but even a more or less regular variation in the length of day – a change in the rate at which the Earth spins. It was Dr R. A. Challinor's report of these effects, in 1971, that put Dr Plagemann and myself on the trail of what we came to call *The Jupiter Effect*; now that we know how the solar wind affects atmospheric circulation it is easy to understand how this in turn affects the rate at which the Earth spins, altering the length of day by a few microseconds over the 11 years of a cycle of solar activity. When air is pushed down towards the equator from high latitudes, for whatever reason, it moves away from the axis of the Earth's rotation and slows it down like a spinning skater with arms extended. When air moves back towards the poles, the opposite effect speeds the Earth's spin up, like a spinning skater with arms folded in.

Seasonal changes in the movement of air masses also change the length of day, up and down by one or two milliseconds in the course of the year. And sudden bursts of solar activity, great flares which send bursts of charged particles across space to interact with the Earth's atmosphere, seem to produce sudden jumps in the length of day, again by amounts of thousandths of a second rather than seconds or minutes. Again, the details of this are given in *The Jupiter Effect* for the interested reader. But the best evidence of all, tying together in one package our present understanding of earthquakes, volcanoes, climate and the rotation of the Earth, comes from the work of Professor Don Anderson, of CalTech.

Anderson has looked at the whole Earth system to find relationships between all of these geophysical phenomena, summarizing a wealth of evidence gathered by many scientists working in different disciplines. He stresses that earthquakes are pretty small-scale phenomena in the global scheme of things, pointing out that typically a seismic disturbance actually releases only about one hundredth or one thousandth of the underlying stress that makes a region prone to earthquake activity in the first place. The list of trigger effects that can set 'quakes off seems to be growing all the time, with the weight of water building up behind dams and the effects of water injected into rock formations, or pumped out, now high on the list of man-made influences on seismicity. But most important of all to our view of the Earth in space is the discovery that there seems to be a global synchronicity in earthquake and volcanic activity outbursts – 'great earthquakes tend to occur in clumps rather than as isolated events', says Anderson, a lesson that was brought home all too clearly in 1976.

These seismic and volcanic 'rhythms' correlate with various other changing features of our planet, including magnetic disturbances, shifts in the poles, and 'wobbles' of the Earth, like the wobbles of a spinning top. And with so many effects interacting, it's hardly surprising that the theorists cannot yet agree on what is cause and what is effect – do the wobbles shake up the seismically active regions to produce earthquakes, or does a burst of large earthquakes shake up the wobbles of the Earth? In all probability, each such effect can stimulate the other, producing a feedback situation, under the right conditions, perhaps rather like the way in which spreading ice cover might squeeze volcanoes into life, producing smoky dust in the atmosphere to act as a sunshield, cooling the Earth further and allowing ice to spread more, thus giving an even bigger squeeze on the volcanoes (see Chapter 4).

Looking again at the whole picture, Anderson draws attention to the many peculiarities that occurred at the beginning of the twentieth century:

> The turn of the century, roughly following the Krakatoa eruption, was a period of major crustal unrest. It was also a period of rapid climatic change, changes in the drift pattern of the magnetic field and changes in the explosive volcanic activity of the earth. Between 1897 and 1914 there were 71 earthquakes of magnitude greater than 8, or roughly 4 per year; of these, 10 were giant earthquakes, magnitude greater than 8·5 . . . To complicate matters, during this same period of time the global mean temperature rose by 1°C, sea level rose, the westward drift of the magnetic field accelerated, the earth's rotation slowed down at an un-

> precedented rate and the Chandler Wobble was growing to a peak value. Unless all of these things are accidentally related it suggests that climatologists, seismologists, tectonophysicists, and geomagnetic students should get together.

And, surely, that they should bring in the astrophysicists as well? For although Anderson has done a remarkable job of bringing together different aspects of the Earth system into a coherent whole, showing clearly that 'earthquakes are just one part of the chain and are more likely to be one symptom rather than either the disease or the cure', he ignores, by and large, the fact that the whole Earth is itself part of a still greater system, the Solar System, and is affected in particular by the vagaries of the Sun's variability. It's just those vagaries, it seems to me, that can explain why 1976 was such an unusual year for seismic disturbances, as well as one of marked climatic variability.

The shakes and shudders of '76

Paradoxically, we can lay the blame for the events of 1976 not at the door of any great outburst of solar activity, but rather on its very quietness, at the calmest period of the solar cycle. It's rather like the classic Sherlock Holmes case in which the vital clue about the boisterous dog is that it did *not* bark in the night! The point is that quite apart from individual outbursts of flaring solar activity there is a more steady pattern of variation in the solar wind. Put crudely, we can imagine the solar wind as blowing more or less steadily in the years between the peaks and troughs of solar activity, gusting more strongly in the years around solar maximum and dying away in the years near solar minimum. The effect on the spin of the Earth – the changing length of day – has been shown in Challinor's work. The change in length of day is greatest when there are *least* sunspots, and least when there are *most* sunspots – the sunspot related effect works to oppose the natural slowing down of the Earth, which goes on all the time, so that the two effects cancel in solar maximum years, and the greatest change is seen in solar minimum years. In terms of a kick or jolt on the spinning Earth, this means that both sunspot minimum and sunspot maximum are important.

Look at it this way. Steady changes in speed never hurt anybody – driving at 100 miles an hour is quite safe on a straight road. But stopping or starting suddenly, or changing direction, can shake you up even at much smaller speeds. At solar maximum, the Earth's change of spin shifts from decreasing to increasing, like stepping on the accelerator after a burst of braking; at solar minimum, the change is from increasing to decreasing spin, like a sudden stab of the brakes following a burst of speed. And both effects jolt up planet Earth. The analogy is not perfect, but it gives the right idea. The conclusions are borne out by Challinor's study, which also showed that earthquakes tend to occur in clusters at both extremes of the solar cycle – solar maximum *and* solar minimum. Since 1976 was exactly at the minimum of the current phase of solar activity, the only surprise, in view of the fact that Challinor's study was published as long ago as 1971, is that anyone should have been caught unawares by the burst of seismic activity.

The whole picture hangs together beautifully. In an era of declining sunspot activity overall, the years of solar minimum, like 1976, give us a taste of events to come, as far as some aspects of climate go – like the average weather of the future, only more so. The circulation of the atmosphere, prodded by the fingers of the solar wind, slips into very low gear, producing the climatic anomalies that we discussed earlier and slowing down the spinning top on which we live quite appreciably. This jolt produces a bigger than usual kick on the seismically active zones of the globe, producing a spate of earthquakes, volcanoes and tidal waves (for so-called 'tidal' waves are in fact the result of underwater earthquakes). It's hardly surprising that even in this scientific age such a variety of disasters has been seen by various people as the work of God or the Devil, or as a sign of the catastrophes which astrology and such prophets as Nostradamus foretell associated with the end of the present world age before the dawning of the Age of Aquarius, the Second Coming of

Christ, or whatever.

It is far more surprising, in fact, that we can see these events as part of a coherent picture linking events on Earth to events elsewhere in our Solar System (which, admittedly, is indeed a form of latter-day astrology). What is more, we can use this new understanding to make firm predictions, including the occurrence of another burst of seismic activity around the time of the next solar maximum, due in the early 1980s. But still it is the longer-term shift in climate that is our main concern at present, and even here study of the changing length of day provides corroboration of the evidence discussed earlier. Since the 1950s, during just the period when we have seen the new climatic patterns beginning to get established, the length of day has been increasing steadily, once the vagaries of the solar cycle effect are smoothed away, at a rate which is now adding a full second to the day each year, according to the regular monitoring carried out at the Royal Greenwich Observatory in England. Just one more global indicator of the fact that we are indeed at the beginning of a new world age – not the Second Coming of Christ, but an age in which food production is likely to be harder than anyone living has yet experienced, while population continues to grow – an age when, to continue the Biblical theme, 'all the plenty shall be forgotten in the land'. Unless, that is, we take appropriate action, on a large scale, in the very near future.

Chapter Nine

Climate and man: the immediate future

Our search for an understanding of the climatic hazards we can expect to face in our own lifetime has taken us far afield. In space, as far as the Sun, the maker of our weather and climate; in time, 10,000 years into the future to predict the arrival of the next Ice Age proper, drawing on evidence from samples 100,000 years old, and older; and across the barriers of scientific specialization to see how climate, earthquakes and changes in the wobble of the terrestrial spinning top are all inter-related. But now, with this broad picture in mind, it is time to come back closer to home, and to look at the likely effects of the climatic changes we now see as inevitable in the next few decades. Several different lines of approach have all pointed the same way – essentially, a return to nineteenth-century climatic conditions, with all that entails, including perhaps more winters like the freezes of 1888 and 1977 in North America.

At a time of increasing global population, with ever more mouths to feed, the impact of any detrimental climatic change becomes of political and personal concern through its effects on agricultural production. In 1977, the political importance of the problem was highlighted in a report by the US Central Intelligence Agency (CIA) meteorologist Russell Ambrosiac, discussed in the journal *Science* by Deborah Shapley. The CIA is particularly concerned about the prospects of reduced yields of grain in the Soviet Union, which might bring the USSR on to the world markets as a regular purchaser of perhaps 20 million tonnes of grain each year – sufficient to have a dramatic impact in the whole structure of the world grain market, and since the US is the major seller on this market, with obvious implications for US foreign policy.

The current shift in climate is seen by Ambrosiac as bringing drier conditions to the regions of European Russia and Kazakhstan which were opened up as massive grain producers in what we now see to have been the unusually good climatic conditions of the 1960s. Climatic change, rather than Nikita Khrushchev's programme of improved technology and fertilizer use, seems to have been the real cause for the great leap forward in Soviet agricultural productivity that occurred during that decade. And if the new pattern – or rather, the return to the old pattern – persists, as seems highly likely, then according to the CIA estimates the USSR will produce only 200 million tonnes of grain, on average, in each of the next five years – and that is 20 million tonnes less than the yearly total called for in the official Five Year Plan, which sees production even rising to 235 million tonnes per year in 1980. In fact, with the dramatic shifts in weather from one extreme to another that now occur, the Soviet grain yields have also fluctuated wildly, from 140 million tonnes in 1975 to 223 million tonnes in 1976, for example.

Other work released by the CIA in the mid-1970s foresees 'mass migrations' and various forms of political instability resulting from a long-term cooling trend of the Northern Hemisphere. This is something of a 'Doomsday Scenario' compared with the more sober, short-term implications of the Ambrosiac report. Such scenarios are nothing new when it comes to making predictions of the future in terms of the food/population problem, which we now see as a more complex food/population/climate/energy problem – the problem of continued growth of our global society. Most recent books about the immediate future have been pessimistic about the food situation – obvious examples are the Ehrlichs' *Population, Resources and Environment*, and the notorious report to the 'Club of Rome', *Limits to Growth*. In the wake of the resultant doom-laden publicity we might naively expect that the climatic problems which we must now confront might be enough to push things 'over the top' and make the situation completely hopeless. In that case, it might well be argued that we might as well enjoy life while we can, since the end is inevitable. But I believe that the true situation is far from being so desperate.

At some of the research centres around the world where 'futurology' is now a major topic, there is a growing awareness that the prophets of doom may have got things substantially wrong. There are problems ahead, to be sure, but they are not insurmountable problems, and by recognizing them in advance we are yet in a position to act effectively to overcome them. The action needed may be on a large scale, involving many nations, and it may imply quite dramatic changes in the way of life we have become used to in the affluent developed nations. But the end of the world is not, after all, at hand – or at least, it need not be. In this context, there is every incentive to continue to work at the solutions to the problems facing us, and no need to bury our heads in the sand. It is vital, however, to make the right decisions soon, especially about food and agriculture, and that is why the new understanding of climate, and the expectation of a return to the conditions of the late nineteenth century, must be used as input to the plans now being laid. But what is the overall framework of the possibilities for the immediate future, within which an understanding of climatic change can make such a significant contribution?

The real 'world food problem'

I have been fortunate in being able to work closely with a team at the Science Policy Research Unit, at the University of Sussex, which is itself deeply involved in the futures debate and has published some of the most significant 'non-doomsday' discussions. This work covers a broad span of social and technological alternatives for the future, including a detailed survey of the prospects for food and agriculture, which is what interests us most here. It turns out that all the doomsday scenarios I have mentioned use as their starting points for projections of the food/agriculture problem a report issued by the US President's Scientific Advisory Committee (PSAC) in 1967, together with US and UN projections of population and demand for food. In fact, more recent studies indicate clearly that there is much more slack in the world food supply system than such studies suggested, while at the same time the potential capacity of the world for agricultural productivity is now known to be greater than was thought in the late 1960s. The result of this revision of old estimates doesn't remove the food problem, but it does suggest a hint of light at the end of the tunnel.

First, and astonishingly enough to anyone only informed by alarmist press reports, during the past two decades of increasing alarm about the world food 'problem' *world population has not been increasing as rapidly as world food production*; reports published by the FAO show an average annual increase in population of 2%, compared with an increase in food production of 2·8% per year. These figures are alone enough to suggest the real 'world food problem', now and in the foreseeable future – not one of production, but of distribution. For, of course, I am not suggesting that people are not starving even now in many parts of the world.

How do we know whether there is enough food in the world to feed the whole population, if it were distributed evenly? The food supply figures are worked out, by such organizations as the FAO and WHO, or PSAC, on the basis of the food available in the 'market place' – that is, food bought and sold on world markets. This takes no account of the fact that many of the poorest people in the world grow food to eat themselves, not to sell, and that this food never gets to market – so such figures always underestimate the availability of food in any year. But still, this underestimate showed, on the basis of the best available data, that in 1970 the available food could have been distributed to provide everyone in the world with a diet supplying 2420 kilocalories per day – and the 1971 FAO/WHO recommendation for an average healthy diet was 2354 kilocalories per person per day. Even the FAO now accepts that people starve because they cannot buy the food that is available, not because there is no food to buy; aid to the less developed nations in recent years has stimulated food production (more than doubled in the past three decades in the Indian sub-continent, for example) but at the same time inequalities have grown, so that the rich eat more but the poorest people eat less.

Using PSAC estimates, we find that the maximum population

the world could support at existing standards of nutrition would be some 7200 million people, while by accepting the lower standard of a so-called '1967 Japanese diet' (that is, the standard fare in Japan in 1967) we could feed even 157,000 million people. In round figures the present population of the world is about 3500 million – so we have a long way to go yet before we run into a technologically insurmountable food problem. Even the proportion of seriously undernourished people in the world today is probably lower than it has ever been, less than 2% of the total population. That still means 60 million people so seriously undernourished that they are using bodily protein reserves (muscles, including heart muscle) as a source of metabolic energy, and is hardly something to be proud of. But the key to this world food problem is now seen to be poverty, not population; it is a political problem, not one of technology.

During the 1950s and 1960s, world production of cereals more than doubled, while the world's population 'only' increased by 50%. But more than half of the extra food went on to the markets of the richest 30% of mankind – the increase was shared almost equally between the 1000 million in the developed countries (DCs) and the 2500 million in the less developed countries (LDCs). So it is not only the poor people *within* nations who cannot buy enough food; the poorer nations themselves buy less food overall, and then that small slice of the global cake is divided unevenly between the inhabitants of the poorer nations, where the differences between richest and poorest individuals are even more pronounced than those between individuals in the DCs. LDCs can only pay for the food they import, under the present world system, with foreign exchange earned by their exports; ironically, agricultural produce makes up three-quarters of these exports, but the LDCs' share of the world market in these products is declining as the DCs produce more and increase their domination of the market. The effect is that in most LDCs the gap between the debts contracted and the ability to repay them is widening from year to year, and the inequalities between rich and poor nations are increasing as well. A dramatic example of the Biblical Matthew effect – the rich get richer while the poor get poorer. As the UK Government's Select Committee on

Overseas Development stressed in a 1976 report, 'the main answer to the world food problem is to give those who are hungry the means to feed themselves, or the income to buy food'.

The kind of effort appropriate to feed the increasing world population has been pin-pointed by, for example, the futures team at Sussex University. Family farms that are worked by the owners of the land, or long-term tenants, generally receive much more attention to detail than large acreages worked on an impersonal business basis. Verges, yards and odd corners may be highly cultivated, with dramatic effects on yields per acre that have already been seen in some parts of the developed world and in some socialist developing countries where enthusiasm and the 'family' approach have been applied on a communal scale – the obvious example is China, and while this does not necessarily mean that the Chinese path is the only hope for the future, it shows clearly that many regions of the 'Third World' are actually capable of supporting two or three times their present population. And all this ignores the possibility of harvesting the oceans, which are certainly under-exploited and badly 'managed' today.

So, far from 'zero growth' being the Holy Grail of planners today, many futurologists now see that a high rate of economic growth, at least in the LDCs, may be the best way to resolve the world food problem. If the problem is one of poverty, then by making the poor nations richer the problem will be resolved – with the side benefit that if past experience is anything to go by, the resulting improvement in living standards will go hand in hand with a decline in birth rate. What do we mean by 'high' growth? The Sussex team has shown that, taking an income level of $1000 per year in the poorest fifth of the population as a target figure to remove the extreme poverty which causes starvation, all that is needed is an annual rate of growth of GNP of between 1 and 4% in Latin America, Europe and southern Africa, with a target date of AD 2000 for the income level to be reached. At similar rates of growth, the poverty trap could disappear from the Middle East, Asia and Africa by 2030 – and no country needs a growth rate bigger than 4% to reach the target by 2050.

What this means is that, other things being equal, the food problem will disappear of its own accord if such rates of growth can be maintained for about half a century. The problem of sustaining such growth for the necessary time is one for the planners, and work by the Sussex team and others has pointed the possible way ahead in more detail than I can here. I am not saying that growth is the only way out of the food/poverty trap; low-growth futures with a deliberate effort to spread the available supplies more evenly would also do the job, if people would be prepared to accept the Robin Hood version of socialism involved. To my mind, a more carefully thought-out version of a growth scenario looks much more likely to prove acceptable to most of the world's population, not least the powerful rich nations, and that is probably the way things will go. The most important factor which is likely to affect progress in this crucial 50–80 year period now facing us (leaving aside such imponderables as global war) is, however, the climatic shift now under way, the return to nineteenth-century conditions. If growth of up to 4% is to be attempted as a solution to the poverty/food problem, it must be borne in mind from the outset that we will be attempting to increase agricultural production at a time when the climate is working against us. Taking a simple example, if we need yields to go up by 4%, but the weather is effectively pushing yields down by 1 or 2% from the peak levels of the 1960s, then we need somehow to make the equivalent of a 6 or 7% increase, in terms of basic agriculture, in order to counteract the effects of the weather. So, in the rest of this chapter I shall be looking at just how much yields are likely to be affected by the return of the old weather pattern.

The political framework

First of all, the events of the past few years show just how dramatic the impact of climate on agriculture can be in the short term. As long ago as October 1973, in testimony before a joint meeting of the US Senate Sub-committee on Foreign Agricultural Policy and the Sub-committee on Agricultural Production,

Marketing and Stabilization of Prices, Professor Reid Bryson drew attention to just this problem. At that time, the evidence for a return to nineteenth-century conditions was not conclusive, but even so the immediate implications of such a shift could already be seen. As Bryson said, if 'we have already returned to a climate like that in the last century . . . this has some very ominous implications for the future', and he pointed out that whereas during the period of benign climate earlier in this century the Indian monsoon failed only once in 18 years, 'now with the deserts moving southwards and the Earth cooling off, the frequency of severe drought in India is starting to pick up again . . . now they have four times as many people to feed as they had at the turn of the century'. In view of the remarkably small amount of progress that has been made in tackling the real causes of the food problem since 1973, and bearing in mind that even on the basis of Bryson's testimony half a decade ago the US winter of 1977, the worst since that of 1888, should have been no real surprise, the following exchange from the transcript of Bryson's testimony should be noted:

> Senator Humphrey: And our country has been derelict in this kind of [climatic] research.
> Dr Bryson: It has, sir, and as a matter of fact, on this point I am not speaking my own opinion only because I understand that the Interagency Co-ordinating Committee on Atmospheric Sciences, the governmental co-ordinating committee of all those units involved in atmospheric sciences, has come to the same conclusion, that the US effort is totally inadequate and that we have to reorder some priorities and put a lot more attention on it because the consequences of doing nothing would be disastrous.

With 1977 having now reminded us just what a return to nineteenth-century conditions really means, at the time of writing (spring 1977) we are still awaiting that reordering of priorities. Small wonder that the problem is seen by the scientists as one of politics rather than of science – with even Bryson stressing in his testimony the political relevance of appropriate action:

> There has been a lot of discussion on establishing a world reserve [of food] under the control of the FAO. I am very concerned about this because of the close linkage between fuel and food, and the imports of fuel and the exports of food.
>
> I would say that one suggestion that ought to be considered is that if we do contribute grain to a world food reserve, we ought to have in return for that a return contribution of five calories of fuel for each calorie of food that we are contributing. That is fair enough because that means that the countries that have fuel but not food can contribute their fair share to the world's supply.

This link between energy and food – directly through mechanization of agriculture, less directly through use of fertilizers derived from soil – is the other side of the 'energy crisis' coin, through which actions by OPEC in raising oil prices to hurt the developed countries rebound on their friends in, say, India by making food production more expensive. The link between energy and food production has been discussed by Gerald Leach and others; any climatic change, such as a cooling, which affects demand for or the supply of energy must also have an effect on food production by modern, energy-intensive agricultural technology, as well as the direct effects on crops of changes in rainfall and length of the growing season. While this link should not be ignored, it is a secondary effect and will not be dealt with at length here. But it is important to realize the political implications of Bryson's comment, very similar, indeed, to statements made by President Carter before his election, to the effect that anyone who wanted an oil war would pretty soon find they were involved in a food war as well.*

* The most apt summing up of the political implications of the food problem seems to me to be that of Professor Walter Orr Roberts, now Program Director for Science, Technology and Humanism at the Aspen Institute in Colorado:

> 'It is human lives that depend on what we provide. What was a $6 billion grain trade in the early 1970s is now around $18 billion and this may rise to $40 billion in another decade. Shall we feed those who walk to our tune, those who are friendly? Shall we feed only the nations that control

Very recently, Stephen Schneider and Richard Temkin have summarized some of the effects of climatic change on food production that occurred in the early 1970s. In 1972, climatic disturbances seriously affected crops in the Soviet Union. A combination of a *lack* of snow (which usually acts as an insulating blanket over the ground through the worst ravages of the Soviet winter) and severe cold in January 1972 led to the loss of a third of the winter wheat; there followed a hot, dry spring which further reduced the winter crop, and also affected the spring wheat. While in other parts of the world the same year saw a late start to the Indian monsoon, droughts in Central Africa, floods in Pakistan and the US mid-West, and a change in Pacific Ocean currents that drastically reduced the anchovy yield for the fishermen off Peru. Overall, global grain production dropped by 1% – sufficient that 'food prices sky-rocketed and food reserves plummeted from roughly two months' supply [for the world population] to about half that much'.

'1973,' say Schneider and Temkin, 'was a record year for world food production. But again in 1974 weather was more highly variable than it had been during much of the previous fifteen-year period.' Again, crop yields were reduced by the weather variability. The same authors also stressed, in their article which was in my hands in late 1976, the importance of weather in affecting energy delivery systems – oil tankers unable to make harbour through ice, trucks and trains snowbound, pipelines and power-lines frozen – a comment which looks like remarkable prophetic ability in the light of the events of the first months of 1977 in the US, but which should, again, have been 'obvious' to anyone who had taken the trouble to listen to reports from the climatologists that all these peculiarities in the weather are most simply understood as a return to nineteenth-century conditions. Further words of wisdom from this team, based at the US National Center for Atmospheric Research in Boulder, Colorado, remind

their population growth? Shall we feed only the nations that have something of value to trade to us, or only the nations that guarantee freedoms we hold dear? Or shall we simply feed the nations that are hungry?'

us of the problems of applying any 'technological fix' to the changing climate. Any strategy planned to çancel out some seemingly harmful change will certainly disturb many other related parts of the complex climatic system, even if we are fortunate enough to affect the target parameter by the right amount – and we just cannot predict, as yet, whether or not these related effects would do more harm than the original, obvious climatic shift. We have to accept, at least for the 50- to 100-year timescale of immediate importance, that deliberate manipulation of climate is not desirable. All we can do – what we *must* do – is to understand how the climate is changing and learn to live with it. Man must adapt to his environment, not the other way around, at least in this important regard.

There is now some hope that the attempt will be made, and in time. The 'winter of '77', as it has already gone down in folklore, was just the timely reminder needed – fortuitously just as a new President was being inaugurated – to make the politicians in the US appreciate the need for a better understanding of the present pattern of climatic change. And in Europe the inauguration of another new President, Roy Jenkins at the head of the European Economic Commission in Brussels, was an equally hopeful sign to the climatologists who have been trying for years to get their message across in the political context. Crispin Tickell, who took up a post as *Chef du Cabinet* with Roy Jenkins when the new President began work in Brussels at the beginning of 1977, is one diplomat who has long been aware of the potential impact of climatic change on agriculture, and just before leaving his previous post with the UK Government's Cabinet Office he chaired a meeting of the British Parliamentary Group for World Government on just this topic. The journal *New Scientist* carried a report of that meeting in its issue of 16 December 1976, spotlighting Tickell's call for a European Centre for Atmospheric Research to be established along the lines of the US establishment NCAR in Boulder, and his proposal that the Common Agricultural Policy of the European Economic Community should be used to build up stockpiles of food as buffers against increased climatic variability, with farmers also being encouraged to grow foods that are at present imported into the EEC. We can only hope that Tickell's views echo those of his boss, and that with this push from the top the European Community may take a lead in showing the world how to cope with the return of climate to nineteenth-century conditions.

Agricultural effects of the new climatic pattern

Just because the 'new' climatic pattern is not really new at all, but a return to conditions that have been common through much of the past millennium, and applied in particular in the nineteenth century, we can tell quite accurately what kind of effects to expect in terms of changes in agricultural productivity and the world food-supply system. To set the scene, we should first appreciate just how little 'spare' food there is in the world today – albeit for political reasons, rather than through practical limitations. Table 9.1 reproduces what Lester Brown has called 'the index of world food security', which shows the changing level of the world's food reserves, indicated in terms of the number of days for which these reserves would suffice to feed the world's population. After a decline through the 1960s, this reserve has now levelled off at about one month's supply; but this conceals the rather disturbing fact that the amount of food needed to keep the world distribution network ticking over – the amount in transit between, say, the grain-rich states of North America and the hungry lands of India and Bangladesh – is also just about one month's supply. In fact, there is now virtually no global reserve of food, in the sense of stockpiles in warehouses being harboured against an emergency. And in this situation, it is a gamble every year on whether the world's millions will be fed. The hand-to-mouth situation we are in is only one step removed from the classic example of bad husbandry, the farmer who eats all his grain and fails even to save the seed needed for next year's crop. Our global husbandry has been so bad that it has brought us to a parlous state in which one bad year could spell death by starvation for tens of millions of people.

Table 9.1 also conceals another interesting piece of information. As the figures show, through the 1960s there was a great

Climatic change and food production
Table 9.1. Index of world food security, 1961 to 1976. Reserves in million metric tons

Year	*Grain*	*Grain equivalent of idle US cropland*	*Total*	*Reserves as days of world grain consumption*
1961	163	68	231	105
1962	176	81	257	105
1963	149	70	219	95
1964	153	70	223	87
1965	147	71	218	91
1966	151	78	229	84
1967	115	51	166	59
1968	144	61	205	71
1969	159	73	232	85
1970	188	71	259	89
1971	168	41	209	71
1972	130	78	208	69
1973	148	24	172	55
1974	108	0	108	33
1975	111	0	111	35
1976*	100	0	100	31

* estimate

stockpile of unused land in the US, land which could have been utilized to grow grain and build up a genuine reserve of food, a reserve which would have been invaluable in the early years of the 1970s, when drought struck so hard across the Sahel, Ethiopia and elsewhere. Why was the land unused? Simply because it was Government policy in the US at that time to pay farmers a subsidy to keep their land idle, for fear that an overproduction of grain might have reduced the price on the world market and made it less economically valuable! If that is the best economic theory the United States has to offer, then surely it is time we tried some other method of planning the economy.

Against this background of bad husbandry and missed opportunity, the chance of even a small decrease in agricultural productivity (remembering also that the world population continues to increase) must be viewed with alarm. And such a decrease is just what Professor Louis Thompson does foresee, as the climate becomes more variable and more like that of the nineteenth century. We have seen that the North American 'breadbasket' is the region of greatest significance as far as any world food surplus is concerned, so it's natural to look more closely at how the climatic change will affect that region. Thompson has calculated the effects of various combinations of changes in rainfall and temperature on the average wheat yields in six states of America, and this provides us with a convenient case study of prospects for the immediate future. The data are gathered together in Table 9.2, and in view of the return to nineteenth-century conditions the numbers to pick out particularly from this table are the ones that apply for a decrease in temperature of either 0·5°C or 1°C, with a look most closely at the figures for decreased rainfall within each of those sections.

Table 9.2 Effects of climatic change in wheat yields in six states of the USA. Average yields under 'standard' climatic conditions are: North Dakota, 25·0; South Dakota, 21·1; Kansas, 25·9; Oklahoma, 25·2; Illinois, 36·3; and Indiana, 36·3 bushels per acre. Figures taken from Louis Thompson, Weather Variability, Climatic Change and Grain Production, *Science*, **188**, p. 535 (1975).

Change in wheat yields (bushel/acre)

Climatic change	*North Dakota*	*South Dakota*	*Kansas*	*Oklahoma*	*Illinois*	*Indiana*
		No change in temperature				
Change in precipitation (%)						
−30	−3·70	−1·85	−2·84	−2·81	+3·08	+3·24
−20	−2·49	−1·58	−1·80	−1·56	+2·25	+2·18
−10	−1·07	−0·67	−0·85	−0·62	+1·22	+1·10
+10	+1·21	+0·42	−0·76	+0·31	−1·41	−1·13
+20	+2·39	+0·60	+1·41	+0·30	−3·02	−2·28
+30	+3·62	+1·42	+1·99	−0·02	−4·84	−3·45

Climatic change	*North Dakota*	*South Dakota*	*Kansas*	*Okla-homa*	*Illinois*	*Indiana*
		No change in precipitation				
Change in temperature (°C)						
−2°	+1·18	+0·47	+1·44	−2·00	+2·36	+1·69
−1°	+0·68	+0·87	+0·74	−0·28	+1·16	+0·88
−0·5°	−0·36	+0·47	+0·37	+0·04	+0·62	+0·44
+0·5°	−0·41	−0·55	−0·38	−0·40	−0·64	−0·46
+1°	−0·86	−1·17	−0·77	−1·16	−1·24	−0·94
+2°	−1·90	−1·64	−1·57	−3·76	−2·69	−1·93
		A decrease of 2°C in temperature				
Change in precipitation (%)						
−30	−2·52	−2·29	−1·40	−4·80	+5·44	+4·93
−20	−3·32	−1·14	−0·35	−3·56	+4·61	+3·88
−10	−0·06	−0·03	+0·59	−2·62	+3·58	+2·80
+10	+2·38	+0·89	+2·30	−1·69	+0·94	+0·57
+20	+4·41	+1·07	+2·86	−1·70	−0·66	−0·58
+30	+4·80	+0·98	+3·43	−2·01	−2·48	−1·76
		A decrease of 1°C in temperature				
Change in precipitation (%)						
−30	−2·62	−1·88	−2·10	−3·09	+4·24	+4·12
−20	−1·81	−0·71	−1·06	−1·83	+3·41	+3·06
−10	−0·56	+0·29	−0·11	−0·90	+2·38	+1·98
+10	+1·89	+2·33	+1·49	+0·04	−0·25	−0·25
+20	+3·07	+1·47	+2·16	+0·03	−1·86	−1·40
+30	+4·30	+1·39	+2·73	−0·29	−3·68	−2·58

Climatic change	*North Dakota*	*South Dakota*	*Kansas*	*Okla-homa*	*Illinois*	*Indiana*
	A decrease of 0·5°C in temperature					
Change in pre-cipitation (%)						
−30	−3·33	−2·29	−2·47	−2·77	+3·70	+3·69
−20	−2·13	−1·11	−1·42	−1·51	+2·87	+2·63
−10	−0·87	−0·20	−0·48	−0·58	+1·84	+1·55
+10	+1·58	+0·80	+1·19	+0·35	−0·79	−0·68
+20	+2·76	+1·07	+1·79	+0·34	−2·39	−1·83
+30	+3·99	+0·98	+2·36	+0·38	−4·22	−3·01
	An increase of 0·5°C in temperature					
Change in pre-cipitation (%)						
−30	−4·11	−3·31	−3·22	−3·21	+2·44	+2·78
−20	−2·90	−2·13	−2·18	−1·96	+1·61	+1·72
−10	−1·48	−1·22	−1·25	−1·02	+0·58	+0·64
+10	+0·63	−0·13	+0·36	−0·09	−2·05	−1·59
+20	+1·98	+0·05	+1·04	−0·10	−3·66	−2·74
+30	+3·21	+0·03	+1·61	−0·42	−5·48	−3·91
	An increase of 1°C in temperature					
Change in pre-cipitation (%)						
−30	−4·56	−3·92	−3·61	−3·98	+1·84	+2·31
−20	−3·35	−2·75	−2·75	−2·72	+1·01	+1·25
−10	−1·93	−4·66	−1·62	−1·79	−0·02	+0·17
+10	+0·19	−0·66	−0·01	−0·85	−2·65	−2·06
+20	+1·53	−0·19	+0·65	−0·85	−4·25	−3·21
+30	+2·77	−0·65	+1·22	+1·18	−6·08	−4·39

Climatic change	*North Dakota*	*South Dakota*	*Kansas*	*Oklahoma*	*Illinois*	*Indiana*
	An increase of 2°C in temperature					
Change in precipitation (%)						
−30	−6·71	−4·39	−4·41	−6·57	+0·39	+1·32
−20	−4·39	−3·22	−3·37	−5·31	−0·44	+0·26
−10	−3·14	−2·39	−2·42	−4·38	−1·47	−0·82
+10	−0·69	−1·13	−0·82	−3·45	−4·10	−3·05
+20	−0·49	−0·67	−0·16	−3·45	−5·71	−4·20
+30	+1·72	−1·12	+0·42	−3·77	−7·53	−5·38

Thompson suggests that if the weather from now until the year 2000 is as variable in the corn and soybean belts of the US as it was at the end of the nineteenth century then the average yield would be reduced by about 3%. That is not too alarming in itself – except that any reduction is alarming when there are no food reserves left! – but that 3% reduction is not likely to be spread evenly across the coming quarter century. Rather, with the increased incidence of extremes of all kinds, we can expect some years to be quite good and others to be very bad indeed, in terms of agricultural productivity. From the evidence of Table 9.2, in the bad years to the end of this century we must expect yields at least 10% below what we have come to think of as 'normal', with the increased variability also bringing bad years more often than during the past quarter century. Of course, in the good years yields could well be greater than 'normal', provided that the best use is made of the available land and that there is no recurrence of a situation in which the farmers are paid *not* to grow crops, for fear of the effect on the market price.

Clearly we need, somehow, to build up that vital reserve of food, an insurance against the bad years, that Stephen Schneider calls for in his book *The Genesis Strategy*, borrowing from the Biblical example of the seven fat years followed by seven lean years. Even that, however, is only a short-term, stop-gap solution to a problem that will not go away, and can only grow bigger, as long as the world economy remains in an unbalanced state where the rich get richer and the poor get poorer. In the short term we can probably muddle through just by saving some of the food grown in the good years; but in the long term we must face up to the world food paradox, that we *can* grow enough food to feed twice the present population of the globe, but even today tens of millions of people are starving. The solutions we need are political, in the broadest sense, not technological; what it boils down to is that the continued survival of our civilization in the face of the climatic threat is assured – provided we have the necessary will to live, and to take appropriate action now.

Chapter Ten

The joker in the pack

The new understanding of the real nature of the climatic threat, and our increasingly confident ability to forecast future climatic trends, rests upon the recognition of the reality of the solar–terrestrial link. With the establishment of the evidence that flickers in the solar furnace do affect the weather here on Earth, we have been able to make climatic forecasts on timescales of decades and centuries even though we are not yet sure of the exact mechanism through which the solar–terrestrial link operates. The decline of solar activity from the unusually high peaks of recent cycles is bound to produce a corresponding return of the weather patterns to the rough and tumble of the nineteenth century – a retreat from the placid plateau which made agriculture so easy in the middle part of the twentieth century. Professor Willett's work, and the studies of Dr Eddy and his colleagues, leave us in no doubt that a cooling comparable to the harsh years at the beginning of the nineteenth century, with more droughts (but also more floods), more freezing winters (but also more scorching summers), and many problems for agriculture will take us up to the end of the present century, and beyond. All this, furthermore, is hitting us at a time when population pressure makes us more vulnerable than ever to crop failures, or even simply poor harvests. To my mind, after years of study of this whole situation, and after having overcome the natural reluctance of an astronomer to believe that our Sun really does flicker, there is no doubt that this is the natural course of events we can expect. But I cannot end without mentioning, if only briefly, the prospect of events taking an unnatural turn.

Everything described in this book so far rests upon one assumption. The climatic changes we have been considering are entirely natural; the joker in the pack is the possibility that man's activities may now be generating new, unnatural influences which may, in the near future, be big enough to disturb the natural patterns of climatic change. Most climatologists who have studied these possible 'anthropogenic' effects agree that man's influence on climate may well be felt quite forcibly, and quite soon – say, in the early part of the next century. Unfortunately, however, there are two strongly opposed schools of thought with a deep division between those theorists who see man's activities hastening a global cooling – perhaps even bringing the 'next' Ice Age a few thousand years early – and those who foresee a man-made global warming, not just cancelling out the natural cooling trend but perhaps going further, melting the ice caps and removing the threat of ice altogether (but replacing it with the threat of water, extensive flooding of low-lying land).

In such a confusing situation, perhaps the best thing is to ignore these various prognostications, and stick by the natural trends, which in any case are going to dominate up to the end of this century, the crucial make or break decades for our civilization. But that is the coward's way out, and I prefer to try to sort out some order from the confusion of the various possible anthropogenic effects, from which we can see how these will most likely influence climate in the long run – beyond the next fifty years or so.

The ways in which man's activities can affect the climate have been summarized recently by Professor Will Kellogg, of the National Center for Atmospheric Research in Boulder, Colorado. Clearing forests and planting crops affects the reflectivity of the Earth (the albedo), and overgrazing in marginal lands reduces the ground cover still further, so that even more heat and sunlight is reflected. Carbon dioxide, released into the air when wood or fossil fuel (oil or coal) is burnt, changes the natural balance by encouraging heat arriving from the Sun and being re-radiated by the ground to be retained – the carbon dioxide absorbs infrared radiation from the ground and acts like a blanket, warming the Earth up (the so-called greenhouse effect). The natural background level of carbon dioxide is estimated as about 290 parts per million of air by volume; already man's activities have raised this to 320 ppmv, and some estimates suggest the concentration will double by the middle of the next century. Such an increase should increase the surface temperature of the globe by between 1½ and 3°C, other things being equal. But alas, other things may not be equal.

Other pollutants, such as the propellant gases from spray cans (chlorofluorocarbons) can produce a greenhouse effect of their own, as well as possibly affecting the ozone balance of the stratosphere and affecting climate less directly in that way. Continued production at the 1974 level would contribute a further 0·5°C of warming within a few decades – but in this case production has already fallen as the potential hazards have been publicized, and in 1975 production was only 80% of the 1974 peak. Oxides of nitrogen from fertilizers may affect climate in similar, but not yet very well understood ways; particles of dust (aerosols) thrown high into the air by man's activities (smoke from factories, wind-blown soil from farmland, and so on) may produce either a warming or a cooling (depending on whose theories you believe); and in the very long term the direct output of heat from power stations and cities will alone be enough to change the nature of climatic changes. The picture could hardly be more daunting. But a semblance of order can be pulled out from the confusion.

The present situation

First, as Professor B. J. Mason, the Director-General of the UK Meteorological Office, stressed in a lecture to the London Royal Society of Arts in 1976, there is no need to panic, since man's activities have not yet had any major impact on climate. Changes of 1°C in the long term mean temperature certainly cannot be ignored, as we have seen. In the UK and North America, such a change represents a change of three weeks in the length of the growing season, and while natural changes of this size have not mattered too much in the past, from now on they are going to be of great significance in view of the inadequate organization of our world food supply and distribution systems to meet the needs of a growing population.

The key question is whether anthropogenic influences are actually acting in the direction of warming or cooling the Earth at present. The cooling effect depends entirely on the possibility that dust particles in the atmosphere act as a kind of sunshield, stopping some of the warmth of the Sun from reaching the ground and imitating the way volcanoes are believed to influence the climate when they erupt and throw dust into the air. For this reason, Professor Reid Bryson has dubbed the effect 'the human volcano' – and he sees this as the main cause of the cooling over the past twenty years which has already taken us one sixth of the way towards a full Ice Age. But others dispute this claim, which rests upon the seemingly straightforward (but in fact controversial) idea that increasing the amount of dust in the atmosphere must cause a global cooling by reducing the amount of heat that reaches the surface of the Earth. Although that ties in with the human experience of the difference between bright, sunny days and those with haze and cloud, such naive extrapolation takes no account of the fact that the dust itself must be absorbing heat. The latest calculations of the effect of *small* dust particles in the atmosphere (aerosols) show that, depending on the exact conditions, such aerosols can cause either a global warming or a global cooling, or even leave the overall heat balance much as it is. The one sure conclusion at present is that the problem cannot be solved by simple approximations.

Larger particles of man-made dust – more similar to the particles that cause cooling after volcanic eruptions – are now removed from industrial effluent before it reaches the atmosphere, and one result of this filtering is that the air over cities is getting cleaner since large polluting particles are no longer released in such large quantities as before; on the other hand, the small particles which get through the filters are being produced in increasing amounts, and these are able to stay in the atmosphere for a long time and spread over wide areas, with the result that the air of countryside and ocean regions away from centres of pollution is becoming dirtier. Writing in the book *The Changing Global Environment*, edited by S. F. Singer, US climatologist Professor Murray Mitchell estimates that man's activities produce, directly or indirectly, about 30% of the total atmospheric load of particles less than 5 microns (5 millionths of a metre) across. Such particles survive in the atmosphere for about nine days, and the total load at any one time is about 40 million tonnes.

In simple terms, the overall effect of such dust particles depends on the nature of the ground beneath. 'Grey' dust over a white background (such as a snowfield) will produce an overall warming, since the dust absorbs heat that would otherwise be reflected away into space. On the other hand, the same dust over a 'black' background (such as farmland) will reflect away some of the heat that would otherwise have reached the ground and been absorbed, so that there is an overall cooling effect. In technical terms, the ground albedo and the back-scattering cross-section of the dust particles are key factors, and in addition an aerosol layer is more likely to produce a net heating if it lies above the usual cloud cover, which provides a 'white' background.

Dr Ruth Reck, of the General Motors Research Laboratories in Warren, Michigan, is one of several researchers who have been studying this problem, and she has looked in particular at the question of just where the balance between heating and cooling can be struck. She finds that for underlying surface albedos greater than 0·6 (that is, 60% of incident radiation reflected) the effect is always a warming, and for albedos less than 0·35 the aerosol layer always produces a cooling influence – and this seems to be independent of the height of the aerosol layer above the surface.

Allowing for the cloud conditions at high altitudes and using a realistic mean aerosol distribution in the computer simulations, it turns out that the present background aerosol density should produce a temperature *increase* of 0·2°C at 85°S, and 0·05°C at 85°N; the difference is caused by the albedo difference of the two hemispheres (0·6 north of 80°N in July; 0·89 south of 80°S in January). This is too small an effect to have been important yet, but suggests that there may be cause for concern in the near future, especially regarding the more sensitive southern polar regions.

The situation is very similar with the other anthropogenic

effect that has been the subject of a great deal of recent debate, the carbon dioxide greenhouse effect. The influence may just be beginning to be felt, but has not yet produced any major impact on the natural patterns of climatic change, as the figures quoted above (and in Table 10.1) show. Drs P. E. Damon and S. M. Kusa, writing in *Science* in 1976, cited scanty evidence of a warming at high southern latitudes which might just be the first sign of the greenhouse effect in action. But they mentioned in their article that this effect should first become apparent at high southern latitudes because the supposed cooling effect of the 'human volcano' would be least there! In fact, as we have just seen, the effect of man-made dust is *greatest* at high southern latitudes and is also a warming effect – so maybe it is the beginnings of a dust-induced warming that are just beginning to show up on instruments in the Antarctic.

And according to another point of view, expressed in *Nature* by Dr R. B. Bacastow in the same year, the increase in carbon dioxide in the atmosphere measured during the 1960s may have been natural, not man made. Bacastow's theory depends on the effect of the Southern Oscillation, a large-scale shift in atmospheric and oceanic circulation patterns that, among other effects, changes the turnover rate of surface waters in the southern oceans. When more ocean water is being turned over at the surface, Bacastow reasons, more carbon dioxide can be absorbed and dissolved, carried away into the depths. When the turnover is slower, the upper layers may become saturated with carbon dioxide and unable to absorb more, leaving a relatively rich lacing in the atmosphere. The very fact that such a theory can explain recent changes in atmospheric carbon dioxide at least as well as the theory that man's activities are producing the change confirms that as yet those anthropogenic influences are small compared with natural fluctuations.

This is even more true of the most direct influence of mankind on climate, the waste heat produced as a kind of thermal pollution. As yet, the total release of heat by all man's activities across the globe amounts to no more than 0·01% of the amount of heat being absorbed from the Sun, and can safely be ignored. Generally accepted theories suggest that a 1% increase in heat would produce a warming by about 2°C overall, but we are a long way from that situation yet. One day – perhaps in a century from now – this problem may become the most important of all anthropogenic influences on climate, but it is not going to affect the problems of those trying to feed the world in the years up to the end of this century.

Table 10.1 Summary of direct anthropogenic influences on the global mean surface temperature (based on Table 2 of W. W. Kellogg's contribution to the book *Climatic Change*)

Effect	*Timescale*	*Influence on temperature*	*Notes*
Increasing carbon dioxide concentration	75 years	increase by 1·5 to 3°C	Almost certain – but must then halt as fossil fuel is exhausted.
Increasing output of fluorocarbons	75 years	increase by 0·5 to 2°C	Unlikely; production already falling from 1974 peak.
Addition of aerosol particles (dust) to atmosphere	starting soon	heating; uncertain amount	
Thermal pollution	125 years	increase by 1 to 4°C	Certain, if our society survives.

Indirect effects

As well as these direct influences on temperature, man's activities are likely to change climate in other ways. The dust particles in the atmosphere could act as nuclei, or seeds, for the formation of rain drops, and radioactive gases released from nuclear power plants can produce ionization of the atmosphere with effects on the processes in thunderstorms which maintain the whole dynamic balance of cloud systems. Most of these possibilities are only vaguely understood at present, and it would be alarmist to take them too seriously. But one indirect effect of man on climate is a real hazard and will serve to provide an example of the need to think before acting as our potential ability to disturb the natural balance of the environment increases.

There is a plan at present to divert some Soviet rivers which now flow north into the Arctic Basin by blasting new channels for them through the Siberian mountains and making them flow south. The purpose of this scheme is twofold, to drain wet regions of western Siberia, and to irrigate dry lands to the south, providing additional much needed farmland. That seems wholly admirable – but there is a real risk that such a move will disturb the balance which maintains the ice cover of the Arctic Ocean, with important consequences on global atmospheric circulation and climate.

The problems have been spelled out by Dr Knut Aagaard and Professor L. K. Coachman, of the University of Washington. They show how a very small change in the salinity of the Arctic Ocean could lead to the disappearance of ice over a wide area, if the fresh water flowing into the basin were removed or reduced. The salinity of the top layers of the Arctic Ocean is much less than in layers a few hundred metres deep, and although fresh water freezes at a higher temperature than salt water, this layering produces a much more important effect, a thermal inversion which keeps the warm water in the depths from reaching the surface and melting the ice. Because of the salt dissolved in the deeper water, it can be considerably warmer than the fresher water at the surface but still more dense; between the surface and 400 m depth salinity produces an *increase* in density 20 times greater than the decrease caused by the corresponding 3°C temperature increase with depth over that range.

The low salinity surface layer is replenished by low salinity ocean water from the North Pacific through the Bering Strait, mixed with fresh water discharged from the chief rivers of Siberia. The rivers provide about 85,000 cubic metres of fresh water every second; if the flow was completely cut off, the salinity layering which keeps the top of the Arctic Ocean cold would almost certainly be destroyed within a decade, the time it takes for the surface waters to circulate around the Arctic Basin. Such a dramatic change is unlikely, but even smaller-scale local changes could have dramatic local effects, which might build up into something of global importance.

The rivers which are the subject of the present diversion scheme discharge now into the southern Eurasian Basin, north from Scandinavia into the Spitsbergen/Franz Josef Land region. There, the freshwater lid on the warmer, more saline waters below is shallower than in most of the Arctic Ocean, and the turnover time for the waters in this local region is only about three years. Aagaard and Coachman suggest that if both the Yenisei and Ob rivers were diverted this lid would disappear from about a million square kilometres of the Eurasian Basin, and that likely results 'would be (1) prolonged ice-free conditions because of deep-reaching free convection . . . (2) the release of large amounts of heat from the warm Atlantic water during the cold months, and (3) the elevation of quite saline water into close proximity with the sea surface'.

The Washington oceanographers draw no conclusions about the desirability of such results, but it is easy to see one possible scenario. With less ice in the north, and warmer waters near the surface, there could be a feedback effect encouraging the break-up and melting of ice over a wide region, with more heat being absorbed by each area of water uncovered than by the highly reflective ice layer it replaces. Any dramatic change in the ice cover of this region would certainly have an effect on the circumpolar circulation and the jet stream, bending the whole pattern of winds and rainfall across north-west Europe and into Siberia. Perhaps such a change would encourage increased snowfall somewhere else at high latitudes, because the warm waters uncovered in the Eurasian Basin would give up water vapour to the air, and this could then fall as snow in cooler regions. This again would affect the overall circulation; we cannot be sure how, but the events of 1976–7 described in Chapter One are an all-too-graphic recent reminder of how susceptible we are to small shifts in this pattern which can become locked in to a feedback state. The climatic effects on regions far removed from the Arctic are completely incalculable as yet, but could be equally significant – it is now widely believed that changes in ice cover at high latitudes are related quite closely to the occurrence of droughts or plentiful rains in the tropics, through the atmospheric circulation systems which link even the monsoons with circulation at high latitudes.

Supporters of such schemes – and this is only one example – argue that because no one can calculate just what harmful effects might result from such activities it is reasonable to proceed with them. The more logical argument is that until the supporters of the scheme can prove that there will be *no* harmful effects it should be shelved!

The longer-term prospects

Unless some critical balance is destroyed by some ill-considered 'experiment' such as the diversion of the Yenisei and Ob rivers, anthropogenic influences on climate are not likely to be felt much in the remaining years of the twentieth century. But from the year 2000 onwards, these effects are likely to be of increasing importance for at least the next century, with most of man's activities acting to produce a warmer Earth. This may, in the long term, be a good thing: a warmer Earth might be a better place to live, but there will be some severe problems of adjustment as some regions warm more than others, and as some areas receive increased rainfall while in others the rainfall declines. There is also the far from insignificant problem of the effects of a rise in sea level, produced by melting ice, on such coastal cities as London and New York, and on the Netherlands and other low-lying countries. Intriguingly, the Earth went through a warm period between 4000 and 8000 years ago – the 'Climatic Optimum' – and this coincided with the development of mankind's civilization. Perhaps, admittedly a slender hope, rather than a return of ice bringing the end of that civilization, we may see an 'artificial' climatic optimum which brings about a renaissance of our global society. In any case, the example of the 'Climatic Optimum' revealed by palaeoclimatic techniques provides us with the best insight into what a slightly warmer Earth is going to be like. Kellogg has summarized the evidence, with an important caution about how far the picture can be justified as a 'forecast':

> We must caution the reader not to accept this as a literal representation of what might occur if the Earth becomes warm again, since the causes and nature of the warming 4000 to 8000 years ago could have been quite different from the nature of society's future effects. While we do not really know what caused that high level of mean temperature to be maintained during the Altithermal [Climatic Optimum], one likely cause is the total output from the Sun, and another possibility is the distribution of sunlight between the northern and southern hemisphere as the Earth's elliptical orbit around the Sun changed . . . the patterns of the general circulation and precipitation would also depend on the mechanism involved.

That said, our hindsight view of the 'Climatic Optimum' is the best available picture of conditions in a warmer Earth. The rainfall distribution of the period, revealed from various studies, is far from being alarming for most agricultural regions of the world today, with wetter conditions being experienced in most of Europe, North Africa and the Middle East, East Africa, India, Western Australia and parts of Asia. Rather disturbingly in view of present global dependence on North American grain, one region known to have been significantly drier when the Earth was warmer is the prairie region of North America, but bearing in mind that each 1°C increase in mean temperature gives an increase in the length of the growing season of roughly ten days at middle and high latitudes, any deficiencies in North American grain output resulting from a global warming of a few degrees should be more than offset by increased production elsewhere.

The greatest changes will be in the coldest regions today, the high latitudes. But there may not be any great melting of the permanent ice sheets. According to Professor M. I. Budyko, a warming of 4°C would be needed to remove the Arctic ice altogether, and the Arctic Ocean has never been entirely free of ice even in the warmest periods of the past million years. Some studies, indeed, show that the East Antarctic ice sheet shrank during the most recent glacial period, and enlarged during the subsequent warming – and each of the ice sheets around the globe must be considered separately, since they respond in different ways to changes in their surroundings.

Each ice sheet is maintained by a balance between snow falling on the top and ice breaking away and melting at the edges. For a slight global warming, the net effect in some cases may be dominated by increased precipitation, so that the volume of the ice sheet grows – this probably explains why in the past the East Antarctic ice sheet, the biggest at present, has developed out of phase with the glaciations in the north. The West Antarctic ice sheet seems to be the most susceptible to any warming at present, and could begin to disintegrate quite fast by geological standards; that is, however, still pretty slow by human standards, and as Kellogg puts it 'obviously one should not expect much action in the time scale of human affairs – that is, for the next few centuries at least'. It is also well worth bearing in mind that ice sheets do not conduct heat very well, and that even if the air at the top of a glacier warms by a couple of degrees, it takes a long time for this heat to penetrate through the ice to the bottom, where it might encourage melting and lubricate the movement of the ice to produce a sudden surge forward. If a surge of the ice occurs in the near future, it will be because of the natural processes set in train over the past thousand years or more, natural processes which we can do nothing to modify now.

Looking beyond the middle of the next century, then, our view is of a warmer Earth in which man's activities counteract the natural cooling trend and tend to prevent the arrival of the 'next' Ice Age. Perhaps we can go further than this. By the time natural changes may be such as to encourage the spread of ice once again in spite of any thermal pollution of man-made greenhouse effect, mankind should be well able to take precautionary measures by dusting the snowfields at high latitudes with soot from fleets of suitably modified aircraft. There seems very little doubt that we could encourage the melting of snow and ice and prevent another full Ice Age developing, albeit at considerable cost and with an effort that would require global, not national, resources. It is rather less likely that in the foreseeable future humankind will be able to induce an Ice Age, either deliberately or inadvertently, and that is a rather comforting thought. As I have suggested, a warmer Earth probably will be a better place to live, although it would be rather different from the present Earth. Not least among these differences will be the political ones – a United States that is a net importer of food grains, as well as raw materials, would hardly be able to carry so much weight in world affairs as it does today, and a Europe with a stable population and increased food production might offer a tempting sight to developing nations in Africa where population could be increasing rapidly enough to offset any benefits to agriculture of a warmer Earth.

A more suitable climate for feeding a lot of people it may be – but the people who are being fed in that better climate will be living in a very different society from our present one.

Immediate prospects

If the longer-term prospect is that, for those who survive to see it, the world may turn out to be a better place to live in, as a result of man's activities, the immediate prospects are much gloomier.

On the timescales of hundreds of years, the natural patterns of climatic change may be broken down, either inadvertently or deliberately, so that the forecasts provided by researchers such as Professor Willett become invalid. The spread of man is, after all, a phenomenon unprecedented in Earth history, and as a result future climatic patterns may no longer echo anything that has happened in the past, including the 'Climatic Optimum' of 4000 to 8000 years ago. The problem is, how do we get through the next few years and decades, in order to have the opportunity to adapt to the new global climatic patterns associated with a warmer Earth?

In the immediate future – next year, and next decade – there is no reason to expect man's activities to affect the direction of the cooling trend, and shift towards increased climatic variability. Just possibly, the first beginnings of the warming influence of mankind will take the worst edge off the climatic decline we are now experiencing – but they will not halt it, let alone return things to the more clement conditions of the earlier part of the twentieth century, until the worst damage has been done. Professor Willett's studies of the solar influence on climate remain the best guide to climate for the rest of this century, and the return to nineteenth-century conditions is surely coming – indeed, it has surely already come, as the droughts, floods and freezes of 1976–7 show.

In the years we are about to live through, humankind must face and overcome many difficulties. The problems of over-population, the inadequacies of the present food distribution situation which gives least to the needy and most to the overfed, pollution, energy and raw materials difficulties all rank high on the list along with the problems of the present climatic deterioration. Understanding climatic change, and acting on that understanding, is just one aspect of the whole complex of problems. If action is indeed taken to ensure adequate supply and distribution of food even in a deteriorating climate, and if then the deterioration turns out to be slightly less extreme than feared – perhaps because of the warming influence of dust and carbon dioxide – the result would be that we might have a little more food than the bare minimum necessary for survival, and that as a result we would be better equipped to tackle other problems on the list. So the prospect of a climatic deterioration slightly less worse than the one outlined earlier in this book is not something to cause complacency, but rather one small straw to clutch at. But we still need to improve our husbandry dramatically.

If we cannot solve the complex of problems, including climatic problems, now facing us then, of course, Professor Willett's longer-term predictions may yet come into their own. For, if society undergoes the more or less complete collapse foreseen by some prophets of doom, there will no longer be the production of carbon dioxide, dust and thermal pollution a hundred years and more from now that seems to be the only way of averting the next Little Ice Age, and the next Ice Age proper. This is the long-term climatic threat – that if society does collapse, the efforts of the survivors to rebuild will be hampered by the harsher climatic conditions that we can expect. To the survivors of a global catastrophe, the conditions of the second part of the twenty-first century, a result of natural climatic changes linked with changes in solar activity, will seem harsh indeed.

So the immediate prospects may hold the key to the future of the whole human race. Whatever happens, it is probably too late to change our bad husbandry sufficiently to avoid a great deal of hardship in the years ahead, and I cannot improve on Professor Kellogg's succinct summary of the situation:

> In the priority list of problems to be faced by society during this transition, long-term change of the average climate probably does not belong at the top. More important, I believe, will be next year's climate, and the year after that. *The famines ahead, wherever and whenever they occur, will assure that millions in the poorer, less developed countries will not survive to witness a 'warmer Earth'*.*

* My italics.

References and bibliography

Bent Aaby, 'Cyclic Climatic Variations in Climate over the Past 5500 years Reflected in Raised Bogs,' *Nature*, **263**, p. 281 (1976).

K. Aagaard and L. K. Coachman, 'Toward an Ice-Free Arctic Ocean,' *EOS*, **56**, p. 484 (1975).

C. G. Abbot, *Smithsonian Miscellaneous Collections*, **448**, p. 7 (1966).

P. H. Abelson (editor), *Food, Politics, Economics, Nutrition and Research*, American Association for the Advancement of Science, *Science* Compendium No. 3 (1975).

Fabian Acker, 'How Safe are British Dams?' *New Scientist*, **73**, p. 315 (10 February 1977).

David Adam, 'Ice Ages and the Thermal Equilibrium of the Earth,' *Journal of Research of the U.S. Geological Survey*, **1**, p. 587 (1973).

David Adam, 'Ice Ages and the Thermal Equilibrium of the Earth II,' *Quaternary Research*, **5**, p. 161 (1975).

J. A. S. Adams, M. S. M. Mantovani and L. L. Lundell, 'Wood versus Fossil Fuel as a Source of Excess Carbon Dioxide in the Atmosphere: A Preliminary Report,' *Science*, **196**, p. 54 (1977).

George Allen, 'Some Aspects of Planning World Food Supplies,' *Journal of Agricultural Economics*, **XXVII**, p. 97 (1976).

D. L. Anderson, 'Frontiers of Knowledge: Earthquakes, Volcanoes, Climate and the Rotation of the Earth.' Paper presented to 56th Annual Meeting of the International Geophysical Union, June 1976.

Anonymous, 'Possible Relationships Between Solar Activity and Meteorological Phenomena,' *EOS*, **55**, p. 524 (1974).

Anonymous (Reuters), '"Storm Age" Feared in Northwest Europe,' article in *NY Herald Tribune*, 26 January 1976.

Anonymous, 'Weather Turns World Economies Topsy-Turvey,' *Business Week*, p. 48, 2 August 1976.

Anonymous, 'The World's Climate is Getting Worse,' *Business Week*, p. 49, 2 August 1976.

Anonymous, 'The Arctic Steadily Becoming Colder,' *Soviet News*, 17 August 1976.

Anonymous, 'Sunspots Indicate Solar Variabilities,' *Astronomy*, **4**, p. 58, October 1976.

Anonymous, 'EEC's Changing Climate,' *New Scientist*, **72**, p. 639, 16 December 1976.

Anonymous, 'Freezing Temperatures,' *New Scientist*, **74**, p. 118, 21 April 1977.

R. B. Bacastow, 'Modulation of Atmospheric Carbon Dioxide by the Southern Oscillation,' *Nature*, **261**, p. 116 (1976).

R. B. Bacastow and C. D. Keeling, 'Atmospheric Carbon Dioxide and Radiocarbon in the Natural Cycle,' in *Carbon and the Biosphere* (ed. G. M. Woodwell and E. V. Pecan), US Atomic Energy Commission CONF 720510 (US National Technical Information Services, Washington, 1973).

Betty Baldwin, James Pollack, Audrey Summers, Owen Toon, Carl Sagan and Warren Van Camp,' Stratospheric Aerosols and Climatic Change,' *Nature*, **263**, p. 551 (1976).

Ian Ball, 'New "Dust Bowl" Fear in US Mid-West,' *Daily Telegraph*, 31 August 1976.

L. A. Barrie, D. M. Whelpdale and R. E. Munn, 'Effects of Anthropogenic Emissions on Climate: A Review of Selected Topics,' *Ambio*, **5**, p. 209 (1976).

E. S. de Beer (ed.), *The Diary of John Evelyn* (Oxford University Press, 1959).

A. Berger, 'Les Variations à longues périodes des éléments de l'orbit terrestre: une question de précision dans la théorie astronomique des paloéoclimats,' *Ciel-Terre*, **91**, p. 261 (1975).

A. Berger, 'Obliquity and Precission for the Last 5,000,000 years,' *Astronomy and Astrophysics*, **51**, p. 127 (1976).

Philip Boffey, AAS Meeting: 'Drought was the Topic of the Week,' *Science*, **195**, p. 964 (1977).

J. R. Bray, 'Glaciation and Solar Activity since the Fifth Century BC and the Solar Cycle,' *Nature*, **220**, p. 672 (1968).

J. R. Bray, 'Cyclic Temperature Oscillations for 0–20, 300 BP,' *Nature*, **237**, p. 277 (1972).

W. S. Broecker, M. Ewing and B. C. Heezen, 'Evidence for an abrupt change in climate close to 11,000 years ago,' *American Journal of Science*, **149**, no. 3679, p. 58 (1960).

C. E. P. Brooks, *Climate Through the Ages* (Benn, London, 2nd ed., 1949).

Lester Brown, 'The World Food Prospect,' *Science*, **190**, p. 1053 (1975).

R. A. Bryson, 'A Perspective on Climatic Change,' *Science*, **184**, p. 753 (1974).

R. A. Bryson, 'World Food Prospects and Climatic Change: Testimony before joint meeting of US Senate Subcommittee on Foreign Agricultural Policy and Subcommittee on Agricultural Production, Marketing and Stabilisation of Prices,' 18 October, 1973.

R. A. Bryson and D. A. Barreis, 'Climatic Change and the Mill Creek Culture of Iowa,' *Journal of the Iowa Archaeological Society*, **15**, p. 1 (1968).

Reid Bryson, 'The Lessons of Climatic History,' *Environmental Conservation*, **2**, p. 163 (1975).

M. I. Budyko, 'The Effect of Solar Radiation Variations on the Climate of the Earth,' *Tellus*, **21**, p. 611 (1969).

M. I. Budyko, *Climate and Life* (Academic Press, New York, 1974).

CIA, *A Study of Climatological Research as it Pertains to Intelligence Problems* (CIA Office of Research and Development, 1974).

CIA, 'Potential Implications of Trends in World Population, Food Production and Climate, OPR-401' (CIA, Washington, 1974).

CIA, USSR: 'The Impact of Recent Climatic Change on Grain Production,' *Research Aid* ER 76-10577U (CIA, Washington, 1976).

N. Calder, 'The Cause of Ice Ages,' *New Scientist*, **72**, p. 576 (1976).

R. A. Challinor, 'Variations in the Rate of Rotation of the Earth,' *Science*, **172**, p. 1022 (1971).

W. J. Chancellor and J. R. Goss, 'Balancing Energy and Food Production, 1975–2000,' *Science*, **192**, p. 213 (1976).

J. Chappell, 'Relationships Between Sea Levels, Oxygen-18 Variations and Orbital Perturbations During the Past 250,000 years,' *Nature*, **252**, p. 1199 (1974).

P. Chýlek and J. A. Coakley, 'Aerosols and Climate,' *Science*, **183**, p. 75 (1974).

C. Clark, *Population Growth and Land Use* (St Martin's Press, New York, 1968).

D. H. Clark, W. H. McCrea and F. R. Stephenson, 'Frequency of Nearby Supernovae and Climatic and Biological Catastrophes,' *Nature*, **265**, p. 318 (1977).

Climate and Food: Climatic Fluctuations and US Agricultural Production. Committee on Climate and Weather Fluctuations and Agricultural Production, Board on Agriculture and Renewable Resources; Commission on Natural Resources, National Research Council (US National Academy of Sciences, 1976).

J. M. Colebrook, 'Possible Solar Control of North Atlantic Oceanic Climate,' *Nature*, **266**, p. 476 (1977).

P. J. Crutzen, I. S. A. Isaksen and G. C. Reid, 'Solar Proton Events: Stratospheric Sources of Nitric Oxide,' *Science*, **189**, p. 457 (1975).

Robert Currie, 'Solar Cycle Signal in Surface Air Temperature,' *Journal of Geophysical Research*, **79**, p. 5657 (1974).

'A.D.,' 'Pouvait on Prevoir la Secheresse en Afrique?' *La Recherche*, **5**, p. 372 (1974).

P. E. Damon and S. M. Kusa, 'Global Cooling?' *Science*, **193**, p. 447 (1976).

W. Dansgaard, S. J. Johnson, N. Reeh, N. Gundestrup, M. B. Clausen and C. U. Hammer, 'Climatic Changes, Norsemen and Modern Man,' *Nature*, **255**, p. 24 (1975).

V. Domingo, D. E. Page and K.-P. Wenzel, 'Measurements of Solar Protons in the Near Earth Magnetotail' in *Correlated Interplanetary and Magnetospheric Observations* (ed. D. E. Page), Reidel, Dordrecht (1974).

N. Eberstadt, 'Myths of the Food Crisis,' *New York Review* (19 February 1976).

John A. Eddy, 'The Maunder Minimum,' *Science*, **192**, p. 1189 (1976).

John A. Eddy, 'The Sun Since the Bronze Age' in *Physics of Solar Planetary Environments* (ed. D. J. Williams), 2, p. 958 (American Geophysical Union, 1976).

A. and P. Ehrlich, *Population, Resources and Environment* (W. H. Freeman, San Francisco, 1970).

R. Eiden and E. Eschelbach, 'Atmospheric Aerosol and its Influence on the Energy Budget of the Atmosphere,' *Zeitschrift fur Geophysik*, **39**, p. 189 (1973).

C. Emiliani, 'Temperatures of Pacific Bottom Waters and Polar Superficial Waters during the Tertiary,' *Science*, **119**, p. 853 (1954).

C. Emiliani, 'Ancient Temperatures,' *Scientific American*, **198**, no. 2, p. 54 (1958).

C. Emiliani, 'Isotopic Palaeotemperatures,' *Science*, **154**, p. 851 (1966).

J. E. Enever, 'Giant Meteor Impact,' *Analog*, **LXXVII**, no. 1, p. 61 (March 1966).

S. Epstein and C. Yapp, 'Isotope Tree Thermometers,' *Nature*, **266**, p. 477 (1977).

FAO, *Indicative World Plan for Food and Agricultural Development* (Rome, 1969).

FAO, *Food and Agricultural Commodity Projections 1970–80* (Rome, 1970).

FAO, *The State of Food and Agriculture* (Rome, 1970).

FAO, *Food and Agricultural Commodity Projections 1975–85* (Rome, 1975).

FAO, *Food and Nutrition*, **1**, nos. 1 and 2 (Rome, 1975).

FAO/WHO, Report of the ad hoc Committee on Energy and Protein Requirements, *Technical Report no. 522* (WHO, 1975).

Rhodes W. Fairbridge, 'Global Climate Change During the 13,500 – b.p. Gothenburg Geomagnetic Excursion,' *Nature*, **265**, p. 430 (1977).

J. C. Farrar, 'Sunspot Weather,' *Farmers' Weekly*, p. xxi, 28 February 1975.

H. Flohn, *Climate and Weather* (Weidenfeld and Nicholson, 1969).

H. Flohn, 'Background of a Geophysical Model of the Initiation of the New Glaciation,' *Quaternary Research*, **4**, p. 385 (1974).

Arthur Giese, 'Stratospheric Pollution, Ultraviolet Radiation and Life,' *Interciencia*, **1**, p. 207 (1976).

J. F. Gillooly and T. G. J. Dyer, 'Structural Relationships Between Corn Yield and Weather,' *Nature*, **265**, p. 434 (1977).

H. Godwin, *Geol. Rundschau*, **40**, p. 153 (1952).

H. Godwin and E. H. Willis, *Nature*, **184**, p. 490 (1959).

Douglas Gough, 'The Shivering Sun Opens its Heart,' *New Scientist*, p. 590, 10 June 1976.

J. S. A. Green, 'The Weather During July 1976: Some Dynamical Considerations of the Drought,' *Weather*, **32**, p. 120 (1977).

John Gribbin, 'Planetary Alignments, Solar Activity and Climatic Change,' *Nature*, **246**, p. 453 (1973).

John Gribbin, 'Climate, the Earth's Rotation and Solar Variations' in *Growth Rhythms and the History of the Earth's Rotation*, ed. G. D. Rosenberg and S. K. Runcorn, p. 413 (Wiley, 1975).

John Gribbin, 'The Ozone Layer,' *New Scientist*, **68**, no. 969, p. 12 (2 October 1975).

John Gribbin, 'Antarctica Leads the Ice Ages,' *New Scientist*, **69**, p. 695, 25 March 1976.

John Gribbin, 'Milankovich Comes In From the Cold,' *New Scientist*, **71**, p. 688 (1976).

John Gribbin, 'Gravity, Dust and Solar Neutrinos,' essay in the 1976 competition of the Gravity Research Foundation, Gloucester, Massachusetts.

John Gribbin, 'Is the Sun a Normal Star?' *Analog*, **XCVII**, p. 32 (February 1977).

John Gribbin and H. H. Lamb, 'Climatic Change in Historical Times' in *Climatic Change* (ed. J. Gribbin), Cambridge University Press (1978).

John Gribbin and Stephen Plagemann, 'Discontinuous Change in Earth's Spin Rate to following Great Solar Storm of August 1971,' *Nature*, **243**, p. 26 (1973).

Peter Gwynne, 'Unnatural Weather Hits US Natural Gas,' *New Scientist*, **73**, p. 318 (10 February 1977).

C. G. A. Harrison and J. M. Prospero, 'Reversals of the Earth's Magnetic Field and Climatic Changes,' *Nature*, **250**, p. 563 (1974).

E. M. Hassan, 'Some Effects of River Regulation on Marginal Seas,' *Ocean Management*, **2**, p. 333 (1975).

J. D. Hays, John Imbrie and N. J. Shackleton, 'Variations in the Earth's Orbit: Pacemaker of the Ice Age,' *Science*, **194**, p. 1121 (1976).

J. R. Hill, 'Long Term Solar Activity Forecasting Using High-Resolution Time Spectral Analysis,' *Nature*, **266**, p. 151 (1977).

C. O. Hines and J. Halevy, 'Reality and Nature of a Sun–Weather Correlation,' *Nature*, **258**, p. 313 (1975).

Arthur Holmes, *Principles of Physical Geology* (Nelson, 1965).

M. Hopkins, H. Scolnik and J. M. McLean, 'Basic Needs, Growth and Redistribution: A Quantitative Approach,' *Working Paper No. 29* (World Employment Programme, ILO, 1975).

David W. Hughes, 'The Inconstant Sun,' *Nature*, **266**, p. 405 (1977).

T. Hughes, 'The West Antarctic Ice Sheet: Instability, Disintegration and Initiation of Ice Ages,' *Reviews of Geophysics and Space Physics*, **13**, p. 502 (1975).

T. Hughes, G. H. Denton and M. G. Grosswald, 'Was There a Late-Würm Arctic Ice Sheet?' *Nature*, **266**, p. 596 (1977).

ICAS, *Report of the Ad Hoc Panel on the Present Interglacial*, ICAS 18b–FY75. (Interdepartmental Committee for Atmospheric Sciences, Federal Council for Science and Technology, National Science Foundation, Washington, 1974).

IFIAS, *The Impact on Man of Climatic Change* (International Federation of Institutes for Advanced Study, Stockholm, 1974).

IMOS, *Fluorocarbons and the Environment*, Report of the Federal Task Force on Inadvertent Modification of the Stratosphere (Council on Environmental Quality, Federal Council for Science and Technology, Washington, 1975).

D. S. Intriligator, 'Evidence of Solar-Cycle Variations in the Solar Wind,' *Astrophysical Journal*, **188**, p. 123 (1974).
L. E. Jacchia, 'Some Thoughts about Randomness,' *Sky and Telescope*, **50**, p. 371 (1975).
R. G. Johnson and B. T. McClure, 'A Model for Northern Hemisphere Continental Ice Sheet Variation,' *Quaternary Research*, **6**, p. 325 (1976).
P. D. Jones (editor), *Climate Monitor* (review of 1976), **5**, no. 1 (Climatic Research Unit, University of East Anglia, Norwich, 1977).
S. Jovčić and J. Young, 'Exceptional European Weather in 1975,' *Weather*, **31**, p. 384 (1976).
L. Joy, *Food and Nutrition Planning*, IDS Reprint no. 107. (Institute of Development Studies, University of Sussex, 1973).
William W. Kellogg, 'Global Influences of Mankind on the Climate' in *Climatic Change* (ed. John Gribbin), Cambridge University Press, 1978.
William W. Kellogg, J. A. Coakley and G. W. Grams, 'Effect of Anthropogenic Aerosols on the Global Climate,' *Proc. WMO/IAMAP Symposium on Long-Term Climatic Fluctuations*, WMO Document 421, p. 323 (WMO, Geneva, 1975).
J. W. King, 'Sun–Weather Relationships,' *Astronautics and Aeronautics*, April 1975, p. 10.
J. W. King, E. Hurst, A. J. Slater, P. A. Smith and B. Tamkin, 'Agriculture and Sunspots,' *Nature*, **252**, p. 2 (1974).
J. A. Kington, 'An Examination of Monthly and Seasonal Extremes Using Historical Weather Maps from 1781: October 1781,' *Weather*, **31**, p. 151 (1976).
J. W. Knight and P. A. Sturrock, 'Solar Activity, Geomagnetic Field and Terrestrial Weather,' *Nature*, **264**, p. 239 (1976).
K. Y. Kondratyev and E. A. Nikolsky, 'Solar Radiation and Solar Activity,' *Quarterly Journal of the Royal Meteorological Society*, **96**, p. 509 (1970).
G. J. Kukla, 'Missing Link Between Milankovich and Climate,' *Nature*, **253**, p. 600 (1975).
John E. Kutzbach, 'The Nature of Climate and Climate Variations,' *Quaternary Research*, **6**, p. 471 (1976).
H. H. Lamb, 'On the Nature of Certain Epochs which Differed

from the Modern (1900–39) Normal' in *Changes of Climate*. Vol. **XX** in the Arid Zone Research Series (UNESCO, 1963)

H. H. Lamb, 'Volcanic Dust in the Atmosphere, with a Chronology and Assessment of its Meteorological Significance,' *Philosophical Transactions of the Royal Society*, **266**, p. 425 (1970).

Gerald Leach, 'Energy and Food Production,' *Food Policy*, **1**, p. 62 (1975).

C. E. Leith, 'The Standard Error of Time-Average Estimates of Climatic Mean,' *Journal of Applied Meteorology*, **12**, p. 1066 (1973).

L. M. Libby and L. J. Pandolfi, 'Climate Periods in Trees, Ice and Tides,' *Nature*, **266**, p. 415 (1977).

L. M. Libby and L. J. Pandolfi, 'Isotope Tree Ring Thermometers,' *Nature*, **266**, p. 478 (1977).

G. W. Lockwood, 'Planetary Brightness Changes: Evidence for Solar Variability,' *Science*, **190**, p. 560 (1975).

W. H. McCrea, 'Ice Ages and the Galaxy,' *Nature*, **255**, p. 607 (1975).

W. H. McCrea, 'Solar System as Space Probe,' *Observatory*, **95**, p. 239 (1975).

J. M. McLean and M. Hopkins, 'Problems of World Food and Agriculture,' *Futures*, **6**, p. 309 (August 1974).

J. D. McQuigg, L. Thompson, S. LeDuch, M. Lockard and E. McKay, *The Influence of Weather and Climate on United States Grain Yields: Bumper Crops or Droughts*, Report to the Associate Administrator for Environmental Monitoring and Prediction (NOAA, US Department of Commerce, 1973).

G. Manley, '1684: The Coldest Winter in the English Instrumental Record,' *Weather*, **30**, p. 382 (1975).

C. G. Markham and D. R. McLain, 'Sea Surface Temperature Related to Rain in Ceará, North-eastern Brazil,' *Nature*, **265**, p. 320 (1977).

B. J. Mason, 'Man's Influence on Weather and Climate,' *Journal of the Royal Society of Arts*, **125**, p. 150 (1977).

J. A. Matthews, '"Little Ice Age" Palaeotemperatures from High Altitude Tree Growth in S. Norway,' *Nature*, **264**, p. 243 (1976).

Samuel W. Matthews, 'What's Happening to Our Climate?' *National Geographic*, **150**, no. 5, p. 576 (November 1976).

A. J. Meadows, 'A Hundred Years of Controversy over Sunspots and Weather,' *Nature*, **256**, p. 95 (1975).

D. H. Meadows, D. L. Meadows, J. Randers and W. W. Behrens III, *Limits to Growth* (Potomac, Washington, 1972).

Georges Michaud, 'Diffusion Timescales and Accretion in the Sun,' *Nature*, **266**, p. 433 (1977).

M. Milankovich, 'Mathematische Klimatetore und Astronomische Theorie der Klimaschwantungen' in *Handbuch der Klimatologie*, ed. W. Köppen and R. Geiger (Borntraeger, Berlin, 1930).

J. Murray Mitchell, Jr., 'An Overview of Climatic Variability and its Causal Mechanisms,' *Quaternary Research*, **6**, p. 481 (1976).

R. M. Morris and R. A. S. Ratcliffe, 'Under the Weather,' *Nature*, **264**, p. 4 (1976).

M. S. Muir, 'Possible Solar Control of North Atlantic Oceanic Climate,' *Nature*, **266**, p. 475 (1977).

A. H. Murphy, A. Gilchrist, W. Häfele, G. Krömer and J. Williams, *The Impact of Waste Heat Release on Simulated Global Climates*, Research Memorandum RM–76–79 (International Institute for Applied Systems Analysis, Laxenburg, Austria, 1976).

NOAA, Committee Report to the National Oceanic and Atmospheric Administration, *The Influence of Weather and Climate on United States Grain Yields: Bumper Crops or Droughts* (US Department of Commerce, Washington DC, 1973).

R. L. Newson, 'Response of a General Circulation Model of the Atmosphere to Removal of the Arctic Ice-Cap,' *Nature*, **241**, p. 39 (1973).

D. Ninkovich and W. L. Donn, 'Explosive Cenozoic Volcanism and Climatic Implications,' *Science*, **194**, p. 899 (1976).

W. D. Nordhaus, *Strategies for the Control of Carbon Dioxide*, Report for the International Institute for Applied Systems Analysis, Laxenburg (Austria) and the Cowles Foundation (Yale, 1975).

R. H. Olson, 'Solar Influences on the Weather,' *Nature*, **253**, p. 686 (1975).

R. H. Olson, W. O. Roberts and C. S. Zerefos, 'Short-term Relationships Between Solar Flares, Geomagnetic Storms, and Tropospheric Vorticity Patterns,' *Nature*, **257**, p. 113 (1975).

R. A. and B. E. Oriti, 'Earthquakes, Moonquakes and Starquakes', Parts I and II, *Griffith Observer*, **40**, no. 10, p. 15 and no. 11, p. 8, October and November 1976.

W. S. B. Paterson, R. M. Koerner, D. Fisher, S. J. Johnson, H. B. Clausen, W. Dansgaard, P. Bucher and H. Oeschger, 'An Oxygen-Isotope Climatic Record from the Devon Island Ice Cap, Arctic Canada,' *Nature*, **266**, p. 508 (1977).

R. R. Payne, 'Are World Protein Supplies Sufficient?' in *Proteins and Human Nutrition* (ed. Porth and Rolls), Academic Press (1973).

M. G. Pearson, 'Snowstorms in Scotland – 1729 to 1830,' *Weather*, **31**, p. 390 (1976).

David A. Perry, 'Oxygen Isotope Ratios in Spruce Cellulose,' *Nature*, **266**, p. 476 (1977).

Gilbert N. Plass, 'Carbon Dioxide and Climate,' *Scientific American Offprints*, no. 823 (1959).

T. M. Prus-Chacinski, 'Comments on the Assumptions Applied in Hydrological Forecasting and Some Remarks on the Droughts in Europe,' *Journal of the Institution of Water Engineers*, **30**, p. 381 (1976).

G. Ramaswamy, 'Sunspot Cycles and Solar Activity Forecasting,' *Nature*, **265**, p. 713 (1977).

S. I. Rasool and J. S. Hogan, 'Ocean Circulation and Climatic Changes,' *American Met. Soc. Bulletin*, **50**, p. 130 (1969).

R. A. S. Ratcliffe, 'The Wet Spell of September–October 1976,' *Weather*, **32**, p. 36 (1977).

Ruth A. Reck, 'Aerosols in the Atmosphere: Calculation of the Critical Absorption/Backscatter Ratio,' *Science*, **186**, p. 1034 (1974).

Ruth A. Reck, 'Aerosols and Polar Temperature Changes,' *Science*, **188**, p. 728 (1975).

G. C. Reid, I. S. A. Isaksen, T. E. Holger and P. J. Crutzen, 'Influence of Ancient Solar Proton Events on the Evolution of Life,' *Nature*, **259**, p. 177 (1976).

Haydon Richards, 'Water Everywhere?' *Nature*, **264**, p. 5 (1976).

Ian Ridpath, 'Flickering Stars Signal New Challenge to Astronomers', *New Scientist*, p. 526, 3 June 1976.

C. Robbins and Jared Ansair, *Profits of Doom* (War on Want, 1976).

Walter Orr Roberts, 'Climate Change and the Quality of Life for the Earth's New Millions,' *Proceedings of the American Philosophical Society*, **120**, p. 230 (1976).

W. O. Roberts and R. H. Olson, 'Great Plains Weather,' *Nature*, **254**, p. 380 (1975).

Howard Rush, Pauline Marstrand, John Gribbin and Gordon MacKerron, 'When Enough is not Enough,' *Nature*, **261**, p. 181 (1976).

Dror S. Sadeh and Meir Meidav, 'Search for Sidereal Periodicity in Earthquake Occurrences,' *Journal of Geophysical Research*, **78**, p. 7709 (1973).

Stephen H. Schneider and Roger D. Dennett, 'Climatic Barriers to Long-Term Energy Growth,' *Ambio*, **4**, no. 2, p. 65 (1975).

S. H. Schneider and R. E. Dickinson, 'Climate Modelling' *Reviews of Geophysics and Space Physics*, **12**, p. 447 (1974).

S. H. Schneider and Clifford Mass, 'Volcanic Dust, Sunspots and Temperature Trends,' *Science*, **190**, p. 741 (1975).

S. H. Schneider and Richard L. Temkin, 'Climatic Changes and Human Affairs' in *Climatic Change* (ed. John Gribbin), Cambridge University Press (1978).

Scientific American, Special Issue on Food, **235**, no. 3 (September 1976).

Ann Sellers and A. J. Meadows, 'Long-Term Variations in the Albedo and Surface Temperature of the Earth,' *Nature*, **254**, p. 44 (1975).

N. Seshagiri, *The Weather Weapon* (National Book Trust, New Delhi, India, 1977).

Deborah Shapley, 'Crops and Climatic Change: USDA's Forecasts Criticised,' *Science*, **193**, p. 1222 (1976).

Deborah Shapley, 'Soviet Grain Harvests: CIA Study Pessimistic on Effects of Weather,' *Science*, **195**, p. 377 (1977).

S. F. Singer (ed.), *The Changing Global Environment* (Reidel, Dordrecht, 1975).

G. L. Siscoe, 'Solar–Terrestrial Relations: Stone Age to Space Age,' *Technology Review*, p. 27, January 1976.

H. P. Sleeper, Jr., 'Planetary Resonances, Bi-Stable Oscillation Modes and Solar Activity Cycles,' *NASA Contractor Report CR–2035* (1972).

SMIC Report *Inadvertent Climate Modification; Report of the Study of Man's Impact on Climate* (MIT Press, 1971).

Peter J. Smith, 'Continental Drift Changes Climate,' *Nature*, **266**, p. 592 (1977).

Charles M. Smythe and John A. Eddy, 'Planetary Tides During the Maunder Sunspot Minimum,' *Nature*, **266**, p. 434 (1977).

SPRU *World Futures: The Great Debate* (eds. C. Freeman and M. Jahodal) Martin Robertson, London (1978).

Edmund Stevens, 'Soviet Scientist Links Moon with Tremors,' *The Times*, 6 December 1976.

Harold Stolov and Ralph Shapiro, 'Investigation of the Responses of the General Circulation at 700mbar to Solar Geomagnetic Disturbance,' *Journal of Geophysical Research*, **79**, p. 2161 (1974).

Minze Stuiver, 'On Climatic Changes,' *Quaternary Research*, **2**, p. 409 (1972).

M. J. Suarez and I. M. Held, 'Modelling Climatic Response to Orbital Parameter Variations,' *Nature*, **263**, p. 461 (1976).

Hans Suess, 'Climatic Changes, Solar Activity, and the Cosmic-Ray Production Rate of Natural Radiocarbon,' *Meteorological Monographs*, **8**, p. 146 (1968).

Don Tarling, 'Continental Drift and Climate,' *Hermes*, January 1974, p. 8.

L. M. Thompson, in *Weather and Our Food Supply*, CAED Report no. 20, Iowa State University (1964).

Louis M. Thompson, 'Weather Variability, Climatic Change and Grain Production,' *Science*, **188**, p. 535 (1975).

Rodney Tibbs, 'Antarctic Research Station Faces Shutdown,' *New Scientist*, **72**, p. 315 (1976).

Crispin Tickell, *Climatic Change and World Affairs*, Center for International Studies, Harvard (1977).

D. G. Trout, 'Effective Temperature and the Hot Spell of June–July 1976,' *Weather*, **32**, p. 67 (1977).

UK Cabinet Office, *Future World Trends* (1976).

US National Research Council, *Climate and Food*. A Report of the Committee on Climate and Weather Fluctuations and Agricultural Production (US National Academy of Sciences, Washington DC, 1976).

A. D. Vernekar, 'Long-Period Global Variations of Incoming Solar Radiation,' *Meteorological Monographs*, **12**, no. 34 (American Meteorological Society, 1972).

Howard Wagstaff, *World Food: A Political Task* (Fabian Research Series, no. 326, 1976).

M. Waldmeier, *The Sunspot-Activity in the Years 1610–1960*. (Schuthess, Zurich, 1961).

W. C. Wang, Y. L. Yung, A. A. Lacis, T. Mo and J. E. Hansen, 'Greenhouse Effects Due to Man-made Perturbations of Trace Gases,' *Science*, **194**, p. 685 (1976).

B. C. Weare, R. L. Temkin and F. M. Snell, 'Aerosol and Climate: Some Further Considerations,' *Science*, **186**, p. 827 (1974).

J. Weertman, 'Milankovich Solar Radiation Variations and Ice Age Ice Sheet Sizes,' *Nature*, **261**, p. 17 (1976).

Harry Wexler, 'Volcanoes and World Climate,' *Scientific American Offprints*, no. 843 (1952).

T. M. L. Wigley, 'Spectral Analysis and the Astronomical Theory of Climatic Change,' *Nature*, **264**, p. 629 (1976).

T. M. L. Wigley and T. C. Atkinson, 'Dry Years in South-east England Since 1698,' *Nature*, **265**, p. 431 (1977).

J. M. Wilcox, 'Solar Activity and the Weather,' *SUIPR Report no. 544* (Stanford University, 1973).

John Wilcox, 'Solar Structure and Terrestrial Weather,' *Science*, **192**, p. 745 (1976).

J. M. Wilcox, P. H. Scherrer, L. Svalgaard, W. O. Roberts, R. H. Olson and R. L. Jenne, 'Influence of Solar Magnetic Sector Structure on Terrestrial Atmospheric Vorticity,' *Journal of Atmospheric Science*, **31**, p. 581 (1974).

J. M. Wilcox, L. Svalgaard and P. M. Scherrer, 'Seasonal Variation and Magnitude of the Solar Sector Structure – Atmospheric Vorticity Effect,' *Nature*, **255**, p. 539 (1975).

Hurd C. Willett, 'The Sun as a Maker of Weather and Climate,' *Technology Review*, p. 47, January 1976.

J. Williams, R. G. Barry, and W. W. Washington, 'Simulation of the Atmospheric Circulation Using the NCAR Global Circulation Model with Ice Age Boundary Conditions,' *Journal of Applied Meteorology*, **13**, p. 305 (1974).

A. T. Wilson, 'Variation in Solar Insolation to the South Polar Region as a Trigger which Induces Instability in the Antarctic Ice Sheet,' *Nature*, **210**, p. 477 (1966).

P. B. Wright, 'Sea Surface Temperature and Climate,' *Nature*, **265**, p. 291 (1977).

A. Zátopek, 'On the Long-Term Microseismic Activity, and Some Related Results,' *Studies in Geodesy and Geophysics*, **19**, p. 14 (1975). (Paper presented to General Assembly of the International Association of Seismology and Physics of the Earth's Interior, Lima, Peru, August 1973.)

A. Zátopek and L. Křivský, 'On the Correlation Between Meteorological Microseisms and Solar Activity,' *Bulletin of the Astronomical Institute of Czechoslovakia*, **25**, p. 257 (1974).

F. E. Zeuner, *Dating the Past* (Methuen, 4th ed., 1958).

Subject Index

Index of Names